NUMBER-CRUNCHING MATH PUZZLES

DICK HESS

An Imprint of Sterling Publishing
387 Park Avenue South
New York, NY 10016

PUZZLEWRIGHT PRESS and the distinctive Puzzlewright Press logo are registered trademarks of Sterling Publishing Co., Inc.

This edition published in 2013.
This book was previously published under the title
All-Star Mathlete Puzzles.

ISBN 978-1-4549-0974-3

Distributed in Canada by Sterling Publishing
℅ Canadian Manda Group, 165 Dufferin Street
Toronto, Ontario, Canada M6K 3H6
Distributed in the United Kingdom by GMC Distribution Services
Castle Place, 166 High Street, Lewes, East Sussex, England BN7 1XU
Distributed in Australia by Capricorn Link (Australia) Pty. Ltd.
P.O. Box 704, Windsor, NSW 2756, Australia

For information about custom editions, special sales, and premium and corporate purchases, please contact Sterling Special Sales at 800-805-5489 or specialsales@sterlingpublishing.com.

Manufactured in the United States of America

2 4 6 8 10 9 7 5 3 1

www.puzzlewright.com

CONTENTS

INTRODUCTION

The puzzles in this volume are primarily for your enjoyment and should be passed on to others for their enjoyment as well. They are meant to challenge your mathematical thinking processes including logical thought, insight, geometrical and analytical thinking, and perseverance. While most of the puzzles will succumb to pencil and paper analysis, there are some that are best tackled with a computer to search for or calculate a solution.

This book is an accumulation of favorite problems I've enjoyed solving over the past 40+ years. I first encountered many of these puzzles in publications that offer problem columns or puzzle sections. These include *Crux Mathematicorum with Mathematical Mayhem*, *Journal of Recreational Mathematics*, *Pi Mu Epsilon Journal*, *Puzzle Corner in Technology Review*, *The Bent*, and *Quantum* (now defunct). The ideas for many others were introduced to me by word of mouth through a delightful community of puzzle solvers.

I owe a huge debt of gratitude to these enthusiasts who love to share their latest challenges and listen to mine. I give special appreciation to Nob Yoshigahara in this regard and echo his sentiments expressed in his book *Puzzles 101*. "Let me thank all who have allowed me to use their ideas, and let the joy of our readers be a reward for all of us who care about and puzzle over hard problems with elegant solutions."

—Dick Hess

CHAPTER 1

THE ELEMENT OF SURPRISE

The puzzles in this section each include an interesting wrinkle, either because of an aspect that is counterintuitive or because there is an elegant element to the solution.

1. THE WATERMELONS

A farmer loads his wagon with 1,000 kilograms of watermelons to take to market. At the start of the trip they are 99% water but partly dry out through evaporation on the trip so that by the end they are 98% water. How much do the watermelons weigh at the end of the trip? Test your intuition by first making a quick guess at the answer, then determine the correct answer.

2. COINS IN THE DARK

There are twenty-six coins lying on a table in a totally dark room. Ten are heads and sixteen are tails. In the dark you cannot feel or see if a coin is heads up or tails up but you may move them or turn any of them over. Separate the coins into two groups so that each group has the same number of coins heads up as the other group. (No tricks are involved.)

3. THE PUNCTURED SEQUENCE

The following sequence has an undisclosed element, x.

... 35, 45, 60, x, 120, 180, 280, 450, 744, 1260, ...

Find a simple continuous function to generate the sequence and compute the surprise answer for x.

4. THE MONSTER TIRE

The world's largest tire (with a radius of 100 miles) is rolling down Broadway at 60 miles per hour. The driver of a vehicle on Broadway first notices the tire approaching from behind when its descending surface is just touching the car's roof directly above her head and 6 feet above the road, thereby pinning it to the road before squashing it flat.

Approximately how long does the driver have to crawl out of the tire's path if she can navigate a crawl space 2 feet high?

5. AQUARIUM PROBLEM

You have a 7-gallon aquarium, a 12-gallon aquarium, and a water supply. The 12-gallon aquarium has dots accurately placed at the centers of its four rectangular side walls. How many steps will it take you to get exactly 8 gallons of water in the larger aquarium if:
(a) the base of the aquarium is a rectangle?
(b) the base is an arbitrary quadrilateral?

6. THE PERSISTENT SNAIL

A snail crawls at one foot per minute along a uniformly stretched elastic band starting at one end. The band is initially 100 feet long and is instantaneously and uniformly stretched an additional 100 feet at the end of each minute while the snail maintains his grip on the band during the instant of each stretch.

At what points in time:
(a) does the snail reach the far end of the band?
(b) does the snail reach a point where its distance from the end of the band after it stretches is shorter than its distance from the end of the band a minute earlier?

7. SHORTCUT GEOMETRY

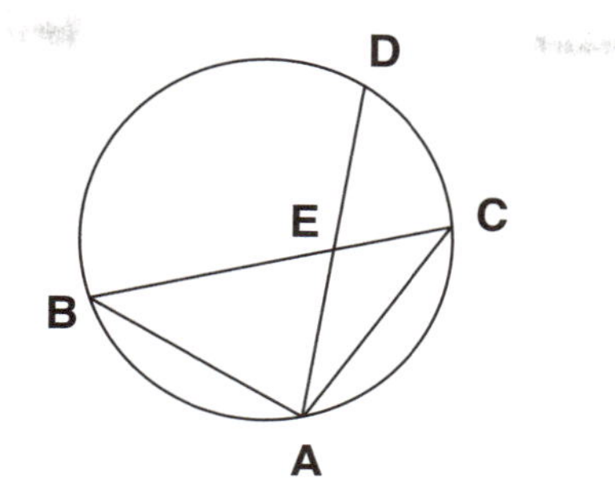

The diagram above appeared in a mathematics test where problems had to be solved under extreme time pressure. AB = AC = 12 and AE = 8. What is the quickest way to determine the length of AD?

8. SHOELACE CLOCKS

In each puzzle you have some shoelaces and matches. The shoelaces burn irregularly like fuses when lit at either end. You are told the burn time for each lace and asked to produce a measured time interval.

(a) Lace 1 burns in 80 minutes. Lace 2 burns in 36 minutes. Produce a 31-minute interval.

(b) Lace 1 burns in 128 minutes. Lace 2 burns in 72 minutes. Lace 3 burns in 48 minutes. Produce a 61-minute interval.

(c) Lace 1 burns in e (= 2.71828…) hours. Lace 2 burns in $\sqrt{2}$ (= 1.4142…) hours. Produce a time interval as close as possible to 1 hour.

(d) In this variation imagine you have an unlimited supply of matches and infinitely fast hands. With two laces, each burning in 60 minutes, produce a 17-minute time interval.

9. MINIMUM HITS

Playing tennis, you have just won one set of singles. What is the least number of times your racket could have struck a ball during the set? (Remember that if a player swings and misses when attempting to serve, it's a fault.)

10. THE THREE SWITCHES

On the ground floor of a building there are three light switches on the wall of the normal kind that show an on or off position. One of them controls a lamp with an ordinary 100-watt light bulb on the third floor of the same building; the other two are not connected to anything, though you have no way of telling this by examining the switches. You are allowed to flip any of the switches as many times as you like before climbing to the third floor only once. Once you are on the third floor you may examine the lamp, but you may not return to the ground floor before identifying which switch controls the bulb.

(a) How do you determine which switch is wired to the lamp?

(b) Now solve the problem if the switches are not marked as to which position is on and which position is off but you know the lamp is initially off.

(c) Solve as in (b) except you don't know whether the lamp is initially on or off.

11. TRANSIT TIME

It is approximately 2244.5 nautical miles from Los Angeles to Honolulu. A boat starts from rest in Los Angeles Harbor and proceeds at 1 knot per hour to Honolulu. How long does it take to arrive there?

12. WHICH CIRCLE IS LARGER?

With a compass, draw a circle on a plane. Without changing the opening of the compass, draw a second circle on a suitably large sphere. Which is larger, the area of the plane enclosed by the first circle or the area of the sphere's surface enclosed by the second circle?

13. MODEST HEXOMINO

Design a tile so that three of them cover as large a fraction as possible of the pictured hexomino. The tiles must not overlap each other or the outer border. One or more of the tiles may be flipped over to achieve the covering.

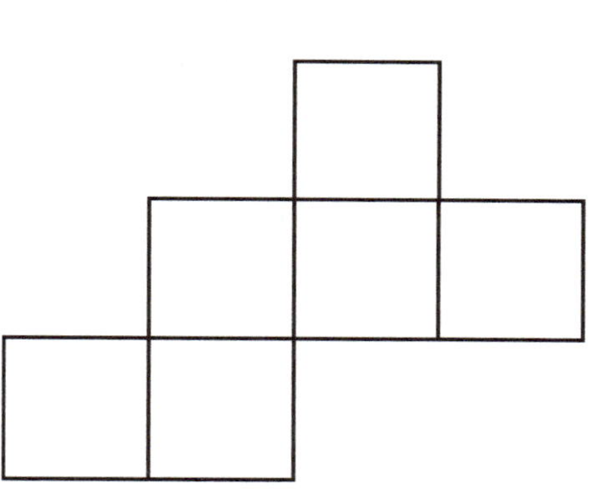

14. LOGICAL QUESTION

Is the area of triangle A, with sides of length 5, 5, and 6, more or less than the area of triangle B, with sides of length 5, 5, and 8?

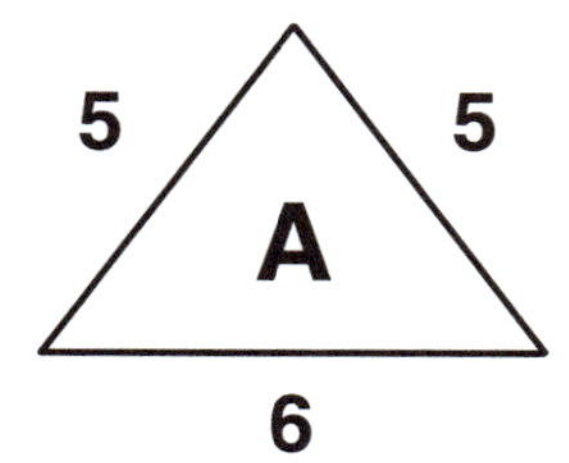

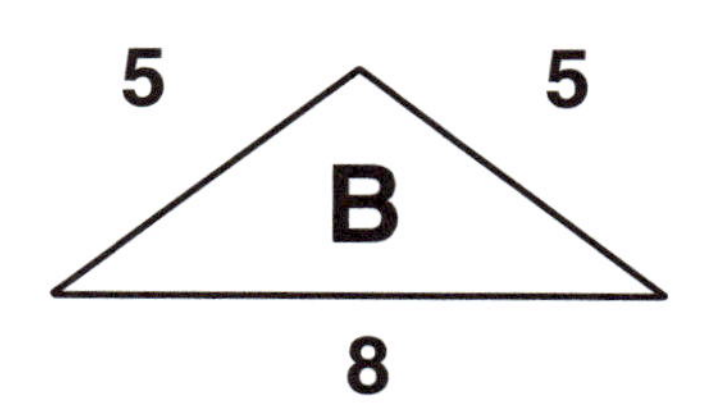

15. THE LONGEST MONTH

What is the longest month of the year in London?

16. OUTSIDE THE BOX

ABCD is a square. AE = 1; BE = 2; CE = 3. Find a clever way to compute the area of the shaded region.

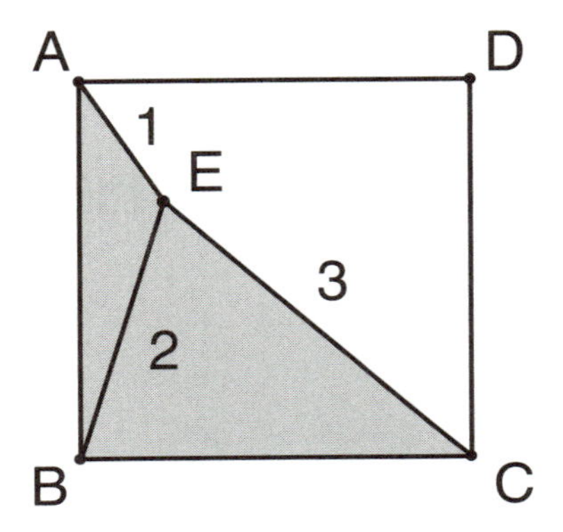

17. HAPPY ANNIVERSARY

At a 20th wedding anniversary party, the hostess tells you that the one of her three children with the lowest age in years has a puzzle he likes her to pose, and proceeds to explain: "I normally ask guests to determine the ages of my three children, given the sum and product of their ages. Since Smith missed the problem tonight and Jones missed it at the party two years ago, I'll let you off the hook." Since you know all three children were born while their parents were married, your response is "No need to tell me more, their ages are ...."

CHAPTER 2

LOGICAL PUZZLES

Put on your thinking cap and tackle these puzzles that will seriously test your powers of logical reasoning.

18. MAGNETIC TAPES

There are 13 magnetic tapes on reels and one empty reel. One can rewind a tape from a full reel to an empty one, thus, reversing its direction. Show how to reverse the direction of every tape while leaving each on its original reel or prove that it is impossible.

19. THE UPSIDE-DOWN WATCH

I recently purchased a 12-hour keychain watch (HH:MM, with hours less than 10 appearing as a single digit) for which right-side-up and upside-down are not well defined. Most of the time it's easy for me to tell which side of the watch should be facing up just by looking at the digits in the display. However, one day I arrived for an appointment thinking I was exactly on time, only to find I was ... well, if I told you how early I was, you'd know when I arrived. When was that? (Assume that digits 0, 1, 2, 5, 6, 8, and 9 read upside-down as 0, 1, 2, 5, 9, 8, and 6, respectively.)

20. TOM, DICK, AND HARRY

You are told that of the three men Tom, Dick, and Harry, there is one Knight (who always tells the truth), one Knave (who always lies), and one Citizen (who may do either at will). They each know who is which. With just two yes-or-no questions, determine who the Citizen is. Then with one more question determine the identity of all three.

21. BLIND LOGIC

There are nine slips of paper with the number 7 on five of them and the number 11 on the other four. Five of the slips go on the hats of logicians A, B, C, D, and E in some order. The other four slips remain hidden. Logician E is blind. Logicians A to D can see the numbers on the others' hats but not on their own. The logicians are error-free in their reasoning and have all the information given so far. They are asked in turn to identify their number.

A: "I don't know my number."

B: "I don't know my number."

C: "I don't know mine."

D: "I don't know mine."

E: "I know my number."

What number is on E?

22. BANANAS AND MONKEYS

There are 14 ladders, each of which has a banana at the top and a monkey at the bottom. A number of ropes connect the rungs of two different ladders such that no rung has more than one rope tied to it. The first monkey climbs up his ladder; each time he encounters a rope, he climbs along it to the other end and continues climbing upward. He is rewarded at the top of the final ladder with a banana. Then the second monkey climbs upward in the same fashion and so on in turn until the fourteenth monkey ends his climb. Show that every monkey ends up with a banana.

23. ODD LOGIC

Two positive odd integers are chosen so that they are either the same or differ by 2. One is written on B's hat, the other on A's hat in such a way that the larger (if there is one) is written on B's hat. The logicians are error-free in their reasoning and have all the information given so far. Their statements are:

A1: "I don't know my number."
B1: "I don't know my number."

A2: "I don't know my number."
B2: "I don't know my number."

A3: "I don't know my number."
B3: "I don't know my number."

A4: "I don't know my number."
B4: "I don't know my number."

And so on, until:

A8: "I don't know my number."
B8: "I now know my number."

What numbers are on A and B?

24. THE CHAMELEONS

On a tropical island live 13 green, 19 brown, and 23 gray chameleons. When two chameleons of different colors meet they each change to the third color. Is it possible that they will eventually all be the same color? Suppose, instead, the initial numbers were 12, 19, and 23?

25. IT'S ALL RELATIVE

A man points to a woman and says "That woman's mother-in-law and my mother-in-law are mother and daughter (in some order)." Find three ways that the man and woman can be related.

26. TENNIS ACES 1

You and I are playing a set of tennis. In the last eight points you have served seven aces and I have served one. What is our score?

27. TENNIS ACES 2

Roddick is playing Federer in the Wimbledon final. Roddick serves aces on six consecutive points during which time Federer does not touch the ball. However, Roddick is still losing the match. What is the precise score in the match?

28. LOGICAL HATS 1

Each of logicians A, B, and C wears a hat with a positive integer on it. The number on one hat is the sum of the numbers on the other two. They see the numbers on the other two hats but not their own. The logicians are error-free in their reasoning and have all the information given so far. They are asked in turn to identify their number.

A: "I don't know my number."

B: "I don't know my number."

C: "My number is 35."

What numbers are on A and B?

29. LOGICAL HATS 2

Positive integers x and y, not necessarily different, are chosen. Their product is written on the hat of logician A and their sum is written on the hat of logician B. Each logician can see the number on the other's hat but not on his or her own. The logicians are error-free in their reasoning and have all the information given so far. After each sees the number on the other's hat one of them says to the other, "There is no way you can know the number on your hat." The other responds, "I now know my number is 136." What numbers are on the hats worn by logicians A and B and who wears which?

30. THE SUM OF SQUARES

Nonnegative integers x and y, not necessarily different, are chosen. Their sum is written on the hat of logician A and the sum of their squares is written on the hat of logician B. Each logician can see the number on the other's hat but not on his or her own. The logicians are error-free in their reasoning and have all the information given so far. They are asked in turn if they are able to identify their numbers. The responses are:

A1: "I don't know my number."
B1: "I don't know my number."

A2: "I don't know my number."
B2: "I don't know my number."

A3: "I now know my number and it is even."

What are the smallest possible values for x and y?

31. BELL-RINGING LOGIC

Each of the three logicians A, B, and C wears a hat with a single integer written on it, from 0 to 8, such that the sum of the three integers is either 6, 7, or 8. A bell rings every so often and after each ring all logicians who know their number announce it and share a prize. The logicians are error-free in their reasoning and have all the information given so far. No one knows his number until after the thirteenth ring, when A announces that he knows his number and wins the entire prize. Can you determine what number appears on each of the three hats?

32. BLIND BELL-RINGING LOGIC

Each of four logicians A, B, C, and D wears a hat with a single integer, from 0 to 8, such that the sum of the four integers is either 6, 7, or 8. Logician B is blind. A bell rings every so often and after each ring all logicians who know their number announce it and share a prize. The logicians are error-free in their reasoning and have all the information given so far. No one knows his number until after the twelfth ring when the blind logician announces and wins the entire prize. What number is on B and what numbers are possible for A, C, and D?

33. TENNIS TOURNAMENT

Nadal, Federer, and Roddick play a challenge tournament in which two play a set and the winner continually stays on court to play the waiting player. At the end of the tournament Nadal has played 15 sets, Federer has played 14 sets, and Roddick has played 9 sets. Who played in set number 13?

CHAPTER 3
GEOMETRIC PUZZLES

And now for a set of puzzles that will test your visual and spatial reasoning abilities (and perhaps your memories of geometry class).

34. OVERLAPPING COOKIES

How are the areas of regions X and Y related?

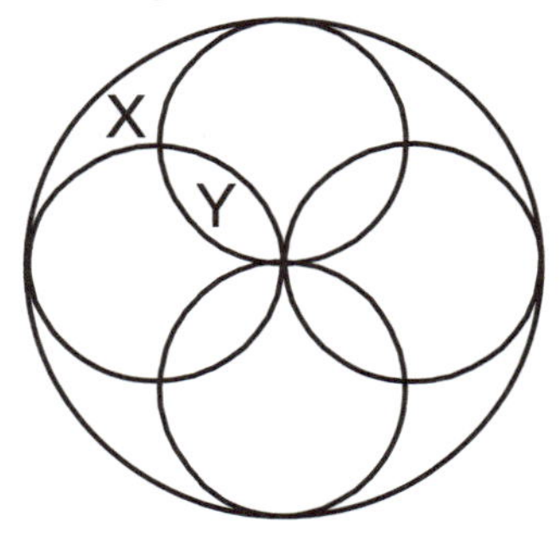

35. MEASURING THE NORTH POLE

Imagine a rubber band stretched along a great circle around the world and over the North Pole as shown below. Given that the rubber band must stretch one extra foot to accommodate it, how tall is the North Pole? (Assume the radius of curvature of the earth at the North Pole is 20,996,000 feet.)

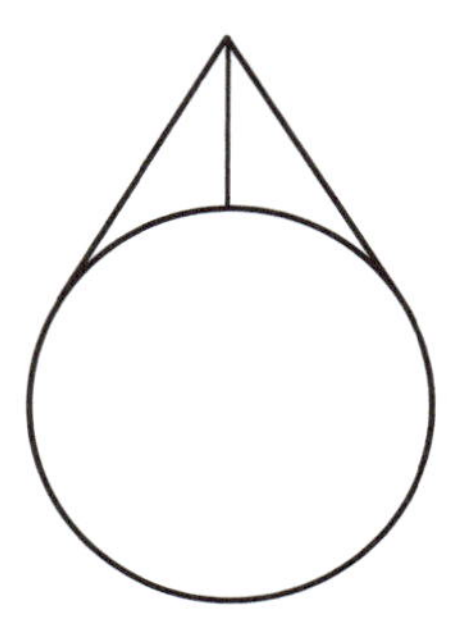

36. TILING THE TRIANGLE

Tile an equilateral triangle with three tiles of identical shape if:

(a) They are all the same size.
(b) They are all of different size.
(c) Two of the tiles are the same size and the third tile is of a different size.

Tiles may be turned over.

37. REFLECTED PYRAMIDS?

Produce two pyramids XABCD and YABCD with the same quadrilateral base ABCD and the altitudes of their eight triangular faces, taken from the vertices X and Y, all equal to 1. Prove or disprove that the line XY must be perpendicular to the plane ABCD.

38. DIVIDE BY THREE

The shape is half a regular hexagon. Cover it with three similar tiles (of the same shape, but possibly different sizes) where two are the same size and one is of a different size. Tiles may be turned over. (Six solutions are known.)

39. DIVIDE BY FOUR

The shape is half a regular hexagon. Cover it with four similar tiles, two of the same size, the other two of different sizes. Tiles may be turned over. (Seven solutions are known.)

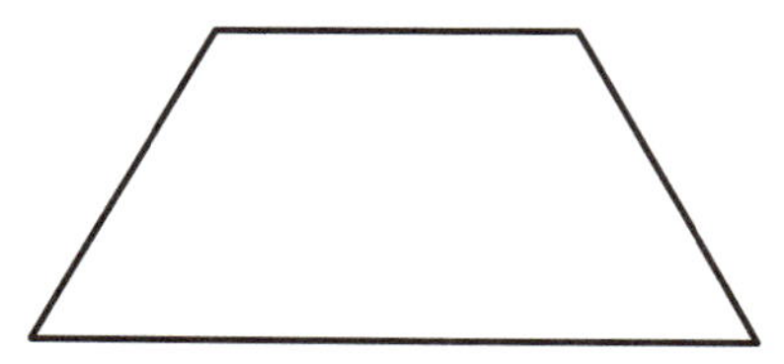

40. SQUEEZE IN

As illustrated below right, it is easy to place 2n unit diameter circles in a $2 \times n$ rectangle. What is the smallest value of n for which you can fit $2n + 1$ such circles into a $2 \times n$ rectangle?

Nob Yoshigahara (in *Puzzles 101*) shows a way to do it when $n = 166$, but a smaller value of n is possible—can you find it?

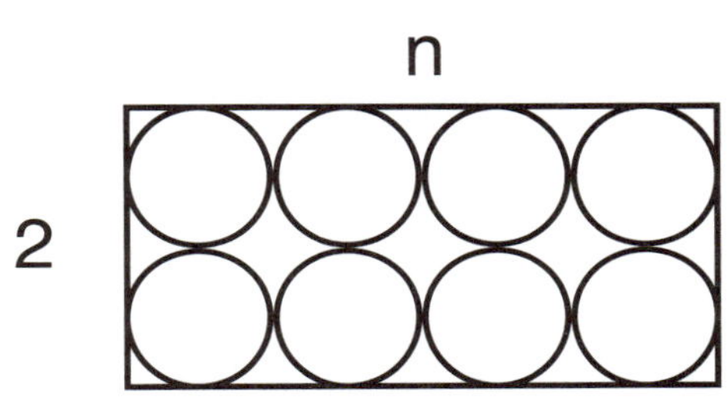

41. THE BILLIARD BALL

A billiard ball with a small black dot on the exact top is resting on a horizontal plane. It is rolled without slipping or twisting so that its contact point with the plane follows a horizontal circle with a radius equal to that of the ball. Where is the black dot when the ball returns to its initial location?

42. SIMILAR TRIANGLES

The figure below shows a 30-60-90 triangle divided into four triangles, all of the same shape. How many ways can you find to do this? 19 solutions are known.

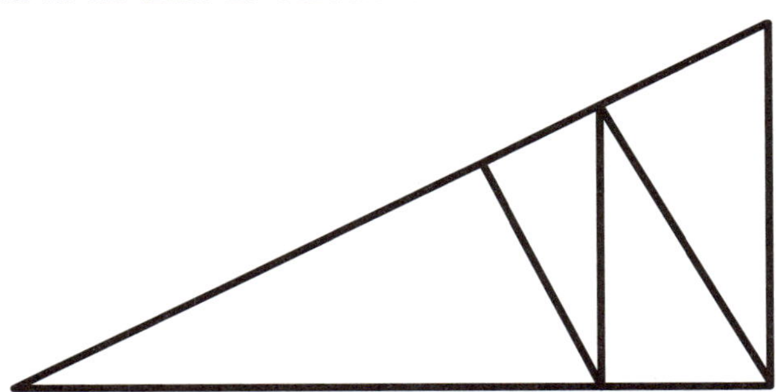

43. MODEST TROMINO

The right tromino can be 100% covered with four congruent tiles as shown in the figure below. Design a tile so that five of them cover over 95% of the right tromino. The tiles must not overlap each other or the outer border. Tiles may be flipped over to achieve the covering.

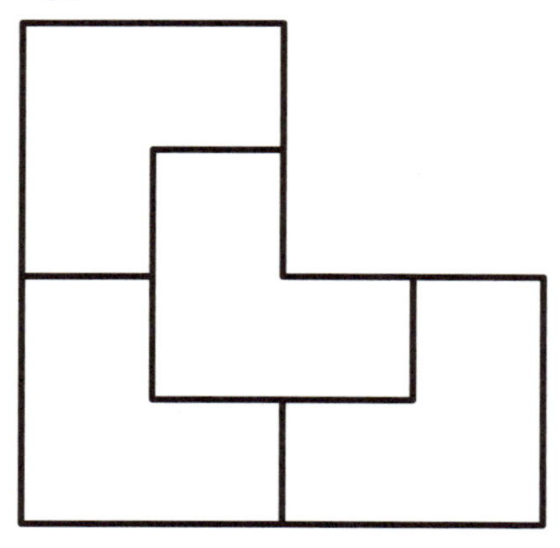

44. MODEST TETROMINO

As shown in the figure below, the L tetromino can be $15/16$ covered with three congruent tiles. Design a tile so that five of them cover over 98% of the L tetromino. The tiles must not overlap each other or the outer border. Any of the tiles may be turned over to achieve the covering.

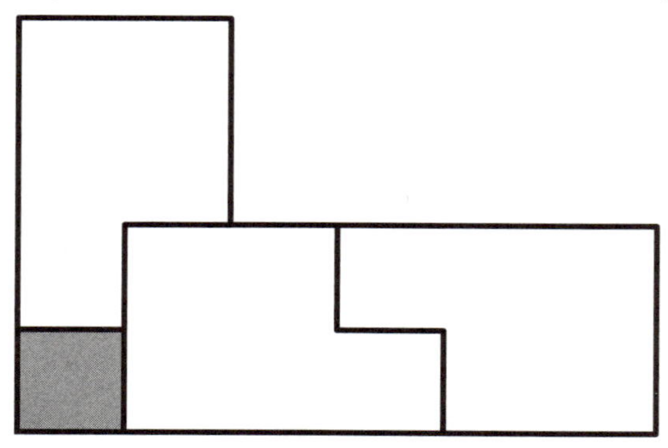

45. SEVEN POINTS

Place seven points at different positions on the plane so that for any three chosen at least two will be exactly one meter apart.

46. WHERE ON EARTH?

A man has breakfast at his camp. He gets up and travels due north along a great circle. After going 10 miles without turning he stops for lunch. After lunch he gets up and travels due north along a great circle. After going 10 miles without turning he finds himself back in camp. Where on earth could he be?

47. LINOLEUM PUZZLE

(a) Cut the large piece below into two pieces that can be reassembled with piece a into an 8×8 square. Pieces may not be flipped. Two solutions are possible.

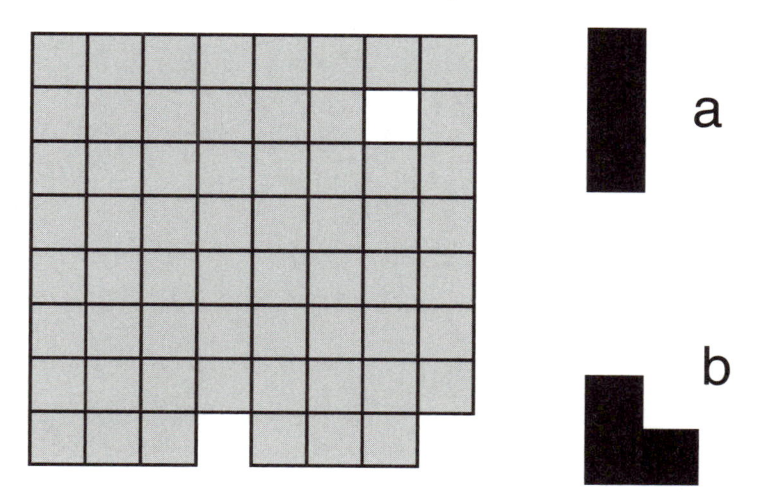

(b) Do the same but use piece b instead of a. Pieces may not be flipped. Two solutions are possible.

48. COUNTING TRIANGLES

List all isosceles triangles, other than right triangles, that use lattice points for all three vertices in:

(a) The L-pentomino (find 12).

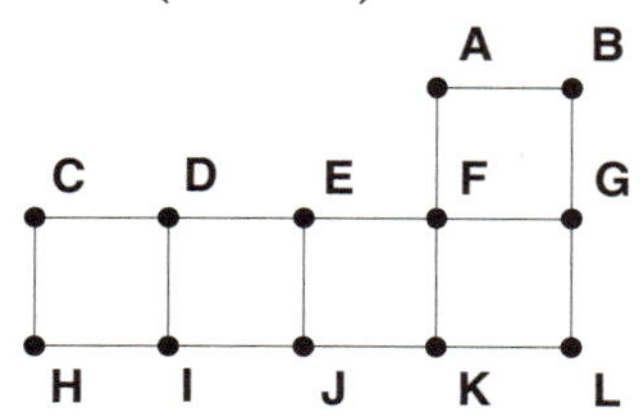

(b) The W-pentomino (find 19).

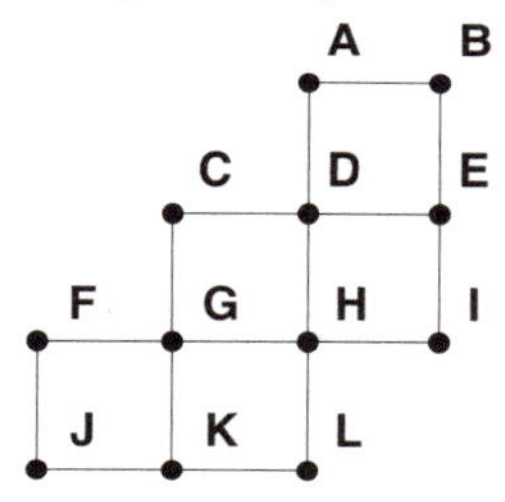

(c) The J-hexomino (find 33).

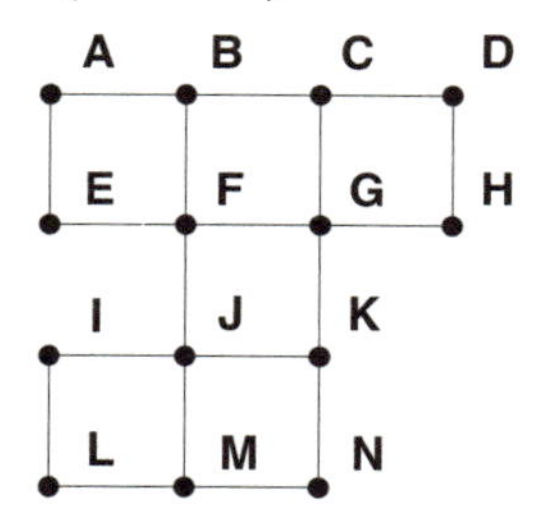

(d) The H-heptomino (find 52).

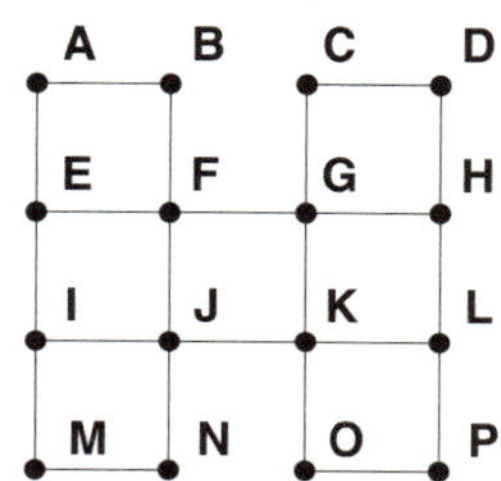

49. DISSECTED SQUARE 1

Dissect a square into similar rectangles with sides in the ratio of 2 to 1 such that no two rectangles are the same size. A solution with nine rectangles is known.

50. DISSECTED SQUARE 2

Dissect a square into similar right triangles with legs in the ratio of 2 to 1 such that no two triangles are the same size. A solution with eight triangles is known.

CHAPTER 4
ANALYTICAL PUZZLES

This is the kind of analysis that doesn't involve an hourly fee. But you can solve on a couch if you like.

51. IRRATIONAL POURING

You are at a lake and have two empty containers capable of holding exactly π (3.14159...) and e (2.7182818...) liters of liquid. How many transfers of liquid will it take you to get a volume of liquid in one container that is within one percent of exactly one liter?

52. WHAT TIME IS IT?

If a clock's second hand is exactly on a second mark and exactly 18 second marks ahead of the hour hand, what time is it?

53. MAXIMUM RATIO

(a) What region inside the unit square has the largest ratio of area to perimeter?

(b) What volume contained inside a unit cube has the largest ratio of volume to surface area? (Answer not known.)

54. COFFEE BREAK

There are two coffee mugs with capacities of 3 and 5 cups, a water supply, and a packet of instant coffee which, when dissolved in one cup of water, constitutes coffee of concentration 100%. Find all positive integer values of n for which it is possible to make coffee of concentration n%.

55. THE KING'S REWARD

"As a reward for your brilliance," said the king, "I give you the land you can walk around in a day. Take some of these stakes with you, pound them into the ground along your way, and be back at your starting point in 24 hours. All the land inside the polygon formed by the stakes will be yours."

(a) If it takes you 1 minute to pound a stake and you walk at a constant speed between stakes, how do you to get as much land as possible?

(b) Would you improve more by walking 2% faster or by taking only 1 second to pound in each stake?

56. THE IRRATIONAL PUNCH

An irrational punch centered on a point P in the plane removes all points from the plane that are an irrational distance from P. What are the least number of placements of the irrational punch on the plane to eliminate all points of the plane?

57. N THIRSTY CUSTOMERS

You have a 37-pint container full of refreshing drink. N thirsty customers arrive, one having an 11-pint container and another having a 2N-pint container. How will you most efficiently measure out 1 pint of drink for each customer to drink in turn and end up with N pints in the 11-pint container and 37 − 2N pints in the 37-pint container if (a) N = 3; (b) N = 5?

58. TRAPPING THE KNIGHT

A knight is placed on an infinite chessboard. If it cannot move to a square previously visited, how can you make it unable to move in as few moves as possible?

59. TENNIS PARADOX

Two evenly matched tennis players are playing a set. The server of any point wins that point with a fixed probability p, where $0 < p < 1$. For what values of p and score situations during a set can the player who is ahead have less than a 50% chance of winning the set?

60. OBTUSE TRIANGLE

Find an obtuse triangle with sides of integer lengths that has acute angles in the ratio of 7/5.

61. MOST UNIFORM DICE

A normal pair of unbiased dice gives a total of 2 through 12 according to the following distribution:

1, 2, 3, 4, 5, 6, 5, 4, 3, 2, 1

How should the spots be changed so as to make the distribution of the total as nearly uniform as possible between 2 and 12? No numbers other than 2 through 12 can occur and you should minimize the sum $(p_2 - 1/11)^2 + (p_3 - 1/11)^2 + \ldots + (p_{12} - 1/11)^2$ where p_k is the probability of rolling k with the two dice.

62. A 9-DIGIT NUMBER

Find a 9-digit number made up of the digits 1, 2, 3, 4, 5, 6, 7, 8, and 9 in some permutation such that when digits are removed one at a time from the right the number remaining is evenly divisible in turn by 8, 7, 6, 5, 4, 3, 2, and 1.

63. THE 1234 QUADRILATERAL

The quadrilateral shown has integer elements a through e. The angles as shown are integer multiples of the smallest.

(a) What is the smallest possible value of c?

(b) What is the smallest possible value of c if α must be obtuse?

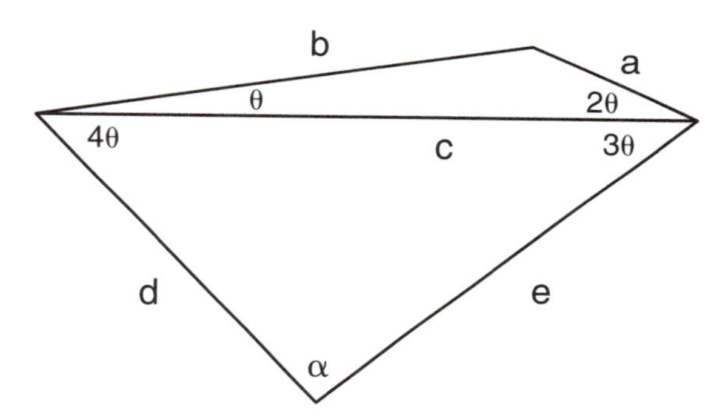

64. DOTS ON THE FOREHEAD

Three people play a game in which an unbiased coin is tossed once for each to determine if they get a red or blue dot on their forehead (red if heads and blue if tails). Each person can see the others' dots but not his own. They are allowed no communication but each must simultaneously guess the color of their own dot or pass. They share a large prize if at least one guesses correctly and no one guesses incorrectly. After hearing the game rules but before the dots are placed the players are allowed a strategy session. What strategy maximizes the chance of sharing the prize?

65. TWO RACERS

Two racers run around a circular track at constant rates. If they run in opposite directions they meet every minute. If they run in the same direction they meet every hour. Find the ratio of their speeds.

66. THE 600 SIXES

Can an integer consisting of 600 digits 6 and any number of digits 0 be the square of another integer?

67. FIND ALL PRIMES

Find all primes p such that $2^p + p^2$ is also a prime. Prove there are no more.

68. THE WANDERING KNIGHT

A chess knight confined to a five-by-five chessboard instantaneously makes a standard knight's move each second in such a way that it is equally likely to move to any of the squares one move away from it. In the long run what fraction of the time does it occupy the center square?

69. THE POND

A pond has a horizontal bottom and vertical sides. To make a platform the owner placed three identical, impermeable cubical blocks side by side resting on the bottom. Placing each block caused the pond's water level to rise. The second and third blocks each caused the level to rise one foot. The first block caused a smaller rise in water level. How high in inches is the water level above the top of the completed platform?

70. COOPERATIVE BRIDGE

You were playing bridge as declarer and held the 4, 3, and 2 of spades, hearts, and diamonds. In clubs you held the 5, 4, 3, and 2. Despite your lack of power you took six tricks in your hand in making a four-club contract. Find three other hands and a line of play from cooperative opponents that allows this to occur.

71. CASCADED PRIME TRIANGLES

The figure below shows two Pythagorean triangles with a common side where 3 of the 5 lengths are prime numbers.

(a) Find other such examples.

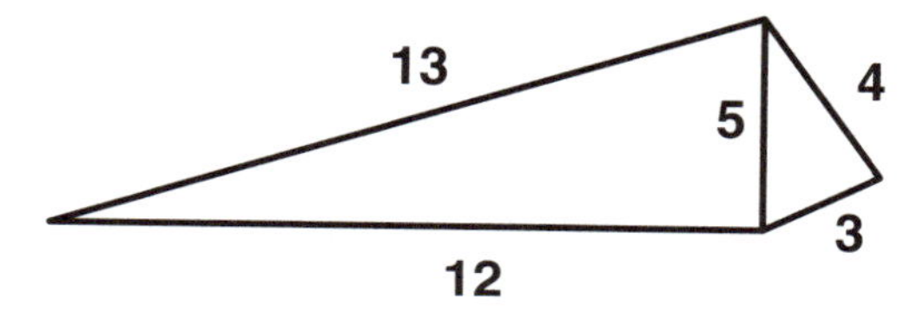

(b) Can a third Pythagorean triangle be abutted as shown below such that 4 of the 7 lengths are primes?

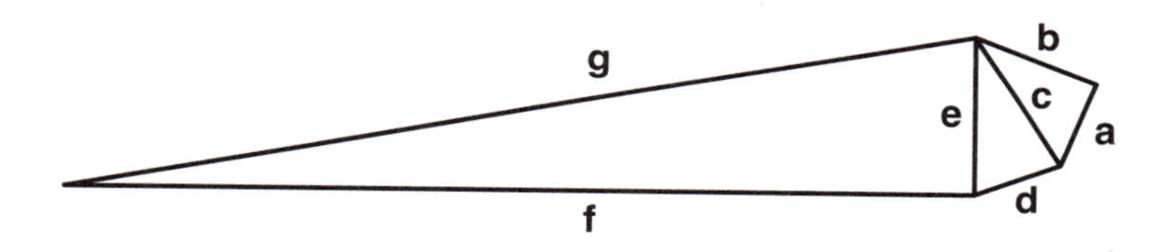

CHAPTER 5
COIN-WEIGHING PUZZLES

In this section each puzzle is a coin-weighing problem in which you are asked to identify all counterfeit coins or a similar task. Coins are indistinguishable except by weight as measured by the equipment provided. One of three types of weighing equipment is used: (1) a simple two-pan balance without weights, (2) a one-pan pointer scale, or (3) a two-pan pointer scale, which shows the signed difference in weight between whatever (if anything) is loaded upon the two pans. Stacks of coins always have identical coins within each stack. Most puzzles are original and are chosen to demonstrate a variety of solving methods.

72. ELEVEN COINS

Among eleven coins, one is known to be counterfeit. With a simple two-pan balance, determine in two weighings if the fake is heavier or lighter than a true coin.

73. FOUR COINS AND FOUR WEIGHINGS

Four coins weigh 4, 5, 6, and 7 grams. In four weighings on a simple two-pan balance, rank the coins by weight.

74. EIGHT COINS

Given: Of eight coins, one is counterfeit.
Conditions: True coins weigh an unknown whole number of grams. The false coin differs in weight from a true coin by less than 4 grams.
Equipment: A one-pan pointer scale.
Weighings: Three.
Objective: Find the fake coin.

75. STACKS OF FIVE COINS

Given: 6 stacks of 5 coins each. One stack has all counterfeit coins.
Conditions: True and false coins are of unknown weights.
Equipment: A two-pan pointer scale that gives the signed difference in weight between whatever (if anything) is loaded on the two pans.
Weighings: Two. Use the fewest coins.
Objective: Find the counterfeit stack.
Variation: Solve for 8 stacks of 5 coins.

76. THREE STACKS OF SIX COINS

You are given three identical-looking stacks of six coins each. Coins in any individual stack all weigh the same amount, and the stacks contain 5-gram, 7-gram, and 12-gram coins.

(a) With a simple two-pan balance determine in two weighings which stack is which.

(b) Repeat with stacks of 6-gram, 7-gram, and 12-gram coins.

77. TEN STACKS OF THREE COINS

Given: Ten identical-looking stacks of three coins. One stack has all counterfeits.
Conditions: True coins weigh an unknown whole number of grams. A false coin differs in weight from a true coin by less than 3.6 grams.
Equipment: A one-pan pointer scale.
Weighings: Two.
Objective: Find the fake stack.

78. N STACKS OF SEVEN COINS

Given: N stacks of seven coins each. One stack has all counterfeits.
Conditions: True coins weigh 14 grams; false coins all weigh either 13 or 15 grams.
Equipment: One-pan pointer scale.
Weighings: Three.
Objective: Find the largest value of N for which you can identify the counterfeits.
Variation: Solve when all false coins each weigh an unknown amount.

79. FOUR STACKS OF TWENTY COINS

Given: Four stacks of twenty coins each.
Coin weights: 1, 2, 3, and 4 grams.
Conditions: Coins in any one stack weigh the same.
Equipment: Simple two-pan balance.
Weighings: Three.
Objective: Find the order of the stacks.
Variation: Coins weigh 1, 2, 6, and 7 grams.

80. N STACKS OF TWELVE COINS

Given: N stacks of twelve coins each. One stack has all light counterfeits.
Conditions: True and false coins are of unknown weights.
Equipment: A two-pan pointer scale which gives the signed difference in weight between whatever (if anything) is loaded on the two pans.
Weighings: Three.
Objective: Find the largest value of N that allows determination of the fakes.
Variation: Solve for four weighings.

81. 2006 COINS

Of a set of 2006 coins, two are counterfeit, one lighter and one heavier than a true coin. With four weighings on a simple two-pan balance determine if the two false coins weigh less than, the same as, or more than two true coins.

CHAPTER 6
MATHDICE PUZZLES

Over 50 years ago a problem called the Four 4's became popular; the challenge was to form mathematical expressions using exactly four 4's to produce specified target values. More recently Sam Ritchie invented the game of MathDice now being sold by ThinkFun. In MathDice, dice are thrown to determine three scoring numbers (1 to 6) and a target number to make with a mathematical expression. Each puzzle in this chapter gives scoring numbers (0 to 9) and a challenge to make a target number according to certain rules. Bring your numerical creativity (and your calculator) for these exercises.

82. FOUR 4'S WARMUP

The Rules: You may use $+$, $-$, $\times$, $\div$, decimal points, and parentheses. No exponents, concatenation (that is, combining two digits into another number; for instance, putting two 4's together to make 44 or .44), roots, factorials, or other mathematical functions.

The Challenge: Use the scoring numbers 4, 4, 4, and 4 once and only once each to form expressions that equal the target numbers 1 to 20. For example, the number 6 can be achieved with the expression $(4 \times .4 + 4 + .4)$. When you solve the challenge, find another example for the number 6.

83. TWENTY MATHDICE CHALLENGES

The Rules: You may use $+$, $-$, $\times$, $\div$, exponents, decimal points, concatenation, and parentheses. No roots, factorials, or other math functions.

The Challenges: Use the scoring numbers once and only once to form expressions that equal the target number. For example, with scoring numbers of 3, 3, and 4 and a target number of 31 we could form $31 = 3^3 + 4 = 34 - 3$. Two expressions are considered the same if one can be immediately derived from the other. For example, $1 \div 2^{-3}$ and 1×2^3, $6 \times .5 \div .3$ and $6 \times 5 \div 3$, and $(6 - 4)^2$ and $(4 - 6)^2$ are pairs of equivalent expressions.

(a) Make 5 using 1, 7, and 7.
(b) Make 20 using 1, 3, and 4.
(c) Make 17 in two ways using 2, 4, and 7.
(d) Make 30 using 2, 7, and 9.
(e) Make 9 in three ways using 2, 5, and 6.
(f) Make 6 using 5, 5, and 9.
(g) Make 4 in two ways using 2, 2, and 7.
(h) Make 20 in two ways using 1, 2, and 5.
(i) Make 12 in four ways using 2, 5, and 8.
(j) Make 16 in two ways using 5, 7, and 8.
(k) Make 15 using 2, 5, and 9.
(l) Make 5 in three ways using 2, 3, and 7.
(m) Make 8 in four ways using 2, 3, and 7.
(n) Make 20 in four ways using 1, 3, and 5.
(o) Make 3 in six ways using 4, 5, and 5.
(p) Make 4 using 2, 3, and 7.
(q) Make 23 in two ways using 5, 5, and 9.
(r) Make 16 in three ways using 3, 5, and 8.
(s) Make 35 in three ways using 1, 2, and 6.
(t) Make 7 in five ways using 4, 5, 9.

84. A DOZEN DEVILISH MATHDICE CHALLENGES

The Rules: As before, you may use $+$, $-$, $\times$, $\div$, exponents, decimal points, concatenation, and parentheses. No roots, factorials, or other math functions.

The Challenges: Use the scoring numbers once and only once to form expressions that equal the target number.

(a) Make 16 using 2, 3, and 4.
(b) Make 25 in three ways using 2, 7, and 9.
(c) Make 11 using 4, 5, and 5.
(d) Make 25 in four ways using 2, 3, and 8.
(e) Make 5 in seven ways using 2, 3, and 4.
(f) Make 25 in three ways using 1, 2, and 3.
(g) Make 16 in three ways using 2, 3, and 8.
(h) Make 3 using 4, 5, and 9.
(i) Make 16 in two ways using 4, 8, and 8.
(j) Make 2 in five ways using 2, 2, and 3.
(k) Make 16 in five ways using 2, 5, and 5.
(l) Make 8 in five ways using 5, 5, and 9.

The remaining MathDice problems in this chapter use expert rules, which include the prior rules expanded to allow repeating decimals, factorials, and roots. For repeating decimals, lines are drawn over digits to indicate they repeat endlessly. So $.\overline{2} = .222222\ldots$, $.\overline{23} = .23232323\ldots$, $.2\overline{3} = .233333\ldots$, and $.\overline{9} = .999999\ldots = 1$. Concatenation, decimal points, and repeating decimals cannot be applied to expressions, only to scoring numbers. The factorial function can only be applied to integers, including expressions that equal an integer. Using the square root function doesn't consume a 2, but other roots consume the digits making up the index of the root. (For instance, expressing a cube root requires the use of the digit 3.) Symbols in an expression may only be used a finite number of times.

85. EXPERT MATHDICE CHALLENGES

The Rules: Expert rules as defined on page 38 (+, −, ×, ÷, exponents, decimal points, concatenation, parentheses, repeating decimals, factorials, and roots).

The Challenges: Use the scoring numbers once and only once to form expressions that equal the target number.

(a) Make 23 using 1, 2, and 6.
(b) Make 21 using 1, 2, and 8.
(c) Make 48 using 2, 2, and 3.
(d) Make 34 using 2, 3, and 3.
(e) Make 31 using 2, 3, and 7.
(f) Make 49 using 2, 5, and 8.
(g) Make 44 using 1, 2, and 4.
(h) Make 44 using 4, 8, and 8.
(i) Make 42 using 4, 8, and 9.
(j) Make 35 in three ways using 4, 9, and 9.
(k) Make 57 using 4, 9, and 9.
(l) Make 23 using 1, 2, and 7.
(m) Make 34 using 2, 3, and 7.
(n) Make 45 using 2, 3, and 7.
(o) Make 22 in two ways using 2, 5, and 8.
(p) Make 26 in three ways using 4, 5, and 5.
(q) Make 38 using 4, 5, and 5.
(r) Make 21 using 4, 8, and 8.
(s) Make 38 using 4, 8, and 8.
(t) Make 48 in six ways using 4, 8, and 8.
(u) Make 55 in two ways using 4, 8, and 9.
(v) Make 44 using 4, 9, and 9.

86. MORE EXPERT MATHDICE CHALLENGES

The Rules: Expert rules as defined on page 38 (+, −, ×, ÷, exponents, decimal points, concatenation, parentheses, repeating decimals, factorials, and roots).

The Challenges: Use the scoring numbers once and only once to form expressions that equal the target number.

(a) Make 55 using 1, 2, and 5.
(b) Make 42 in two ways using 2, 2, and 7.
(c) Make 23 using 2, 2, and 8.
(d) Make 33 using 2, 3, and 8.
(e) Make 35 using 2, 3, and 8.
(f) Make 66 using 0, 1, and 4.
(g) Make 31 using 2, 7, and 9.
(h) Make 33 using 4, 5, and 5.
(i) Make 43 in two ways using 4, 5, and 5.
(j) Make 46 using 4, 5, and 5.
(k) Make 39 in two ways using 4, 8, and 9.
(l) Make 46 using 4, 8, and 9.
(m) Make 61 in two ways using 4, 9, and 9.
(n) Make 44 using 5, 7, and 8.
(o) Make 19 using 1, 2, and 3.
(p) Make 50 in two ways using 1, 2, and 3.
(q) Make 38 using 1, 2, and 5.
(r) Make 39 using 1, 2, and 5.
(s) Make 31 using 1, 2, and 7.
(t) Make 34 in two ways using 2, 4, and 7.
(u) Make 34 in two ways using 2, 5, and 6.
(v) Make 26 using 3, 5, and 8.
(w) Make 35 in two ways using 3, 5, and 8.
(x) Make 41 using 3, 5, and 8.
(y) Make 42 in two ways using 4, 5, and 5.
(z) Make 29 using 4, 8, and 8.
(aa) Make 23 in two ways using 5, 7, and 8.
(bb) Make 36 using 5, 7, and 8.

87. MAKE 28

The Rules: Expert rules as defined on page 38 (+, −, ×, ÷, exponents, decimal points, concatenation, parentheses, repeating decimals, factorials, and roots).

The Challenges: Use the scoring numbers once and only once to form expressions that equal 28.

(a) Make 28 using 1, 2, and 6.
(b) Make 28 in two ways using 1, 3, and 5.
(c) Make 28 using 1, 5, and 9.
(d) Make 28 using 1, 6, and 7.
(e) Make 28 using 1, 7, and 7.
(f) Make 28 using 2, 2, and 8.
(g) Make 28 using 2, 3, and 3.
(h) Make 28 using 2, 5, and 8.
(i) Make 28 using 5, 6, and 6.
(j) Make 28 in three ways using 5, 6, and 7.
(k) Make 28 in three ways using 5, 6, and 9.
(l) Make 28 in two ways using 6, 8, and 9.

88. THE THREE 4'S PROBLEM

The Rules: Expert rules as defined on page 38 (+, −, ×, ÷, exponents, decimal points, concatenation, parentheses, repeating decimals, factorials, and roots).

The Challenges: Use the scoring numbers 4, 4, and 4 once and only once each to form expressions that equal the target numbers.

(a) Make 19.
(b) Make 34 (3 ways).
(c) Make 35.
(d) Make 45 (2 ways).
(e) Make 53.
(f) Make 55 (2 ways).
(g) Make 59.
(h) Make 63.
(i) Make 68 (2 ways).
(j) Make 78.

89. KILLER MATHDICE CHALLENGES

The Rules: Expert rules as defined on page 38 (+, −, ×, ÷, exponents, decimal points, concatenation, parentheses, repeating decimals, factorials, and roots).

The Challenges: Use the scoring numbers once and only once to form expressions that equal the target number.

(a) Make 88 in two ways using 1, 2, and 3.
(b) Make 75 in two ways using 1, 2, and 4.
(c) Make 33 in two ways using 1, 2, and 7.
(d) Make 27 in six ways using 2, 2, and 5.
(e) Make 23 in five ways using 2, 2, and 9.
(f) Make 40 using 2, 4, and 7.
(g) Make 52 using 4, 5, and 9.
(h) Make 19 using 4, 8, and 8.
(i) Make 26 in four ways using 4, 8, and 8.
(j) Make 45 using 4, 8, and 8.
(k) Make 66 in three ways using 4, 9, and 9.
(l) Make 33 using 1, 2, and 2.
(m) Make 29 in four ways using 1, 2, and 4.
(n) Make 33 in three ways using 1, 2, and 4.
(o) Make 45 using 1, 2, and 6.
(p) Make 39 using 1, 7, and 8.
(q) Make 29 in three ways using 2, 2, and 3.
(r) Make 57 using 2, 3, and 4.
(s) Make 38 in two ways using 4, 9, and 9.
(t) Make 56 in six ways using 4, 9, and 9.
(u) Make 21 in two ways using 5, 5, and 9.
(v) Make 25 in four ways using 5, 7, and 8.

CHAPTER 7
PUZZLES THAT WEAVE A STORY

These final five puzzles are included because they each involve an amusingly told story and also provide a good puzzle challenge. Enjoy.

90. MINING DECISION

The planet Alpha Lyra IV is an oblate spheroid. Its axis of rotation coincides with the spheroid's small axis, just as one observes for Sol III. Its internal structure, however, is unique, being made of a number of regions bounded by coaxial right circular cylinders. Each region is of homogeneous composition. The common axis of these cylinders is the planet's rotational axis. The outermost of these regions is nearly pure krypton and the next inner region is an anhydrous fromage. Cosmic, Inc. is contemplating mining the outer region, and the company's financial planners have found that the venture will be sound if there are more than one million cubic spandrals of krypton in that region. (A spandral is the Lyran unit of length.) Unfortunately, the little that is known about the krypton region was received via sub-etherial communication from a Venerian pioneer immediately prior to the Venerian's demise. The pioneer reported with typical Venerian obscurantism that the ratio of the volume of the smallest sphere that could contain the planet to the volume of the largest sphere that could be contained within the planet is 1.331 to 1. It also reported that the straight line distance between it and its copod was 120 spandrals. (A copod corresponds roughly to something between a sibling and a root shoot.) The reporting Venerian was mildly comforted because the distance to its copod was the minimum possible distance between the two. By nature the

Venerians can only survive on the planet's surface at the krypton/fromage boundary and by tragic mistake the two copods had been landed on disconnected branches of the curve of intersection of the krypton/fromage boundary and the planetary surface. Should the venture be undertaken?

91. EXCAVATING ON TAURUS

Taurus, a moon of Alpha Lyra IV, was occupied by a race of knife-makers eons ago. Before they were wiped out by a permeous accretion of Pfister-gas they dug a series of channels in the satellite surface. A curious feature of these channels is that each is a complete and perfect circle, lying along the intersection of a plane with the satellite's surface. An even more curious fact is that Taurus is a torus (doughnut shape). Five students of Taurus and its ancient culture were discussing their field work one day when the following facts were brought to light: The first student had dug the entire length of one of the channels in search of ancient daggers. He found nothing but the fact that that the length of the channel was 30π spandrals. The second student was very tired from his work. He had dug the entire length of a longer channel but never crossed the path of the first dagger digger. The third student had explored a channel 50π spandrals in length, crossing the channel of the haggard dagger digger. The fourth student, a rather lazy fellow (a laggard dagger digger?), had merely walked the 60π-spandral length of another channel, swearing at the difficulties he had in crossing the channel of the haggard dagger digger. The fifth student, a rather boastful sort, was also tired because he had thoroughly dug the entire length of the largest possible channel. How long was the channel that the braggart haggard dagger digger dug?

92. LIFE AND DEATH ON ALPHA LYRA III

The inhabitants of Alpha Lyra III recognize special years when their age in Lyran years is of the form $a = p^2q$, where p and q are different prime numbers. The first few such special years are 12, 18, and 20. On Alpha Lyra III one is a student until first reaching a special year immediately following a special year; one then becomes a master until first reaching a year that is the third in a row of consecutive special years; next one becomes a sage until death, which occurs in the first special year that is the fourth in a row of consecutive special years. (a) When does one become a master? (b) When does one become a sage? (c) How long do the Lyrans live? (d) Do five special years ever occur consecutively? (e) Do six special years ever occur consecutively?

93. HUMPTY DUMPTY

"You don't like arithmetic? I don't very much, either," said Humpty Dumpty. "But I thought you were good at sums," said Alice. "So I am," said Humpty Dumpty. "Good at sums; oh certainly. But what has that got to do with liking them? When I qualified as a Good Egg—many, many years ago, that was—I got a better score in arithmetic than any of the others who qualified. Not that that's saying a lot. None of us did as well in arithmetic as in any other subject." "How many subjects were there?" asked Alice, interested. "Ah!" said Humpty Dumpty, "I must think. The number of subjects was one-third of the top score obtainable in any one subject. And I ought to mention that in no two subjects did I get the same score, and that is also true of the others who qualified as Good Eggs." "But you haven't told me ..." began Alice. "I know I haven't told you the total score one had to obtain to qualify. Well, I'll tell you now. It was a number equal to four times the maximum score obtainable in one subject. And we all just barely managed to qualify." "But how many ..." said Alice. "I'm coming to that," said Humpty Dumpty. "How many of us were there? Well, when I tell you that no two of us obtained the same assortment of scores—a thing which was only just possible—you'll be well on the way to the answer. But to make it as easy as I can for you, I'll put it another way. The number of others who qualified when I did, multiplied by the number of subjects (I've told you about that already), gives a product equal to half the total score obtained by each Good Egg. And now you can find out all you want to know." He composed himself for a nap. Alice was almost in tears. "I can't," she said, "do any of it. Is it differential equations, or something I've never learned?" Humpty Dumpty opened one eye. "Don't be a fool, child," he said crossly. "Anyone ought to be able to do it who is able to count on five fingers." What was Humpty Dumpty's score in arithmetic?

94. THE KNIGHTS OF RIGHTEOUSNESS

The Knights of Righteousness (117 in all) are regrettably imprisoned by a powerful king. They occupy individual, isolated cells and are given the following notice by the king's representative. "As you know, you all face execution, but the king, in the spirit of mercy, is giving you a chance to go free. One at a time you will each enter The Room of Judgment, which contains 117 boxes numbered on the outside from 1 to 117. We have written your 117 names on 117 cards—one per card—shuffled them and put one card in each box. During your turn in The Room of Judgment you may open no more than 98 boxes to gain access to the 98 cards inside and see if your name is among them. Then you must leave the room exactly as you found it and return to your isolation cell without contact or communication with anyone. The Good News: If each knight finds the card with his name on it then you will all go free. The Bad

News: If any knight fails to find the card with his name on it then you will all be executed." The king (who knows a little mathematics) has cleverly designed things so that if the knights choose boxes randomly their survival chances are less than one part in a billion. This is true despite the probability of any one knight's finding his name is 83.76%. Upon reading this notice the chief knight sends a hasty message to the king: "Please allow us knights 30 minutes to be together and pray as is our custom before engaging in a challenge." The king (in all his mercy) grants the knight's request. The leader of the knights (who knows even more mathematics than the king) has a plan to increase the knight's odds by conferring before facing The Room of Judgment. What should he do and to what level can he increase the survival chances of The Knights of Righteousness?

SOLUTIONS

1. The Watermelons

They weigh exactly 500 kilograms. Initially there are 10 kilograms of material that is not water. That doesn't evaporate, so in the end there are still 10 kilograms of material that is not water. In the beginning this 10 kilograms accounts for 1% of the load. After the trip it accounts for 2% of the load. Thus the final load weighs exactly 500 kilograms.

2. Coins in the Dark

Separate the coins into group A with 10 coins and group B with 16 coins. If there are x heads in group A then there will be $10 - x$ heads in group B. Turn over all coins in group A and each will have $10 - x$ coins facing heads up.

3. The Punctured Sequence

The sequence can be expressed by the simple continuous function, $F(n) = 120(2^n - 1) \div n$ when n is not equal to 0. The terms shown are for $n = -3, -2, -1, 0, 1, 2, \ldots 6$. To get x take the limit of F(n) as n goes to 0. Therefore, $x = 120\ln 2 = 83.17766\ldots$

4. The Monster Tire

In the picture $r = 100$ miles, $u = 2$ feet, $h = 6$ feet, and the tire moves at $v = 88$ ft./sec. $a^2 = r^2 - (r - u)^2 = 2ru - u^2$; $b^2 = 2rh - h^2$. Time $= x \div v = (b - a) \div v = 12.089356$ seconds. The driver should have time to escape.

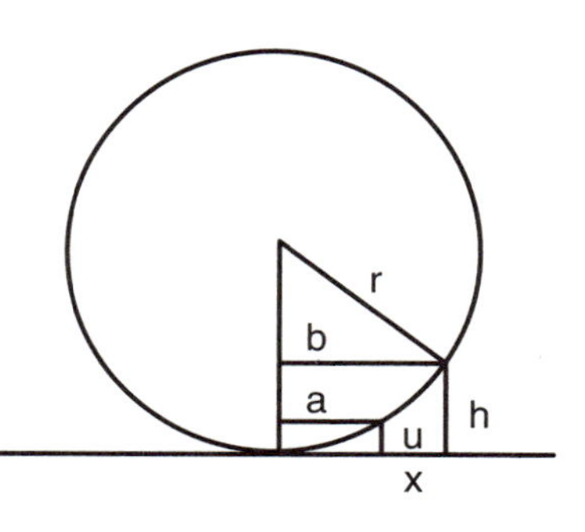

5. Aquarium Problem

It can be done in one step. The figure shows an aerial view of the aquarium with upper rim ABCD. Diagonals of ABCD cross at K with crossing angle θ. Points M and N are center points of two sides of the aquarium. Fill and tip the 12-gallon aquarium appropriately until the water level passes through a top corner (A) and center dots on the two far sides (M and N).

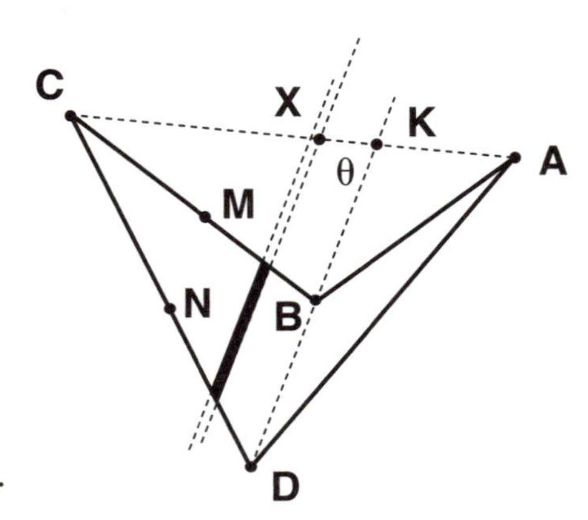

Let X be an arbitrary point along AC and let the height of the aquarium be H. Define W=BD, L=AC , a=AK, b=KC, and x=AX. When the aquarium is full the volume of water in it is (aHWsinθ) ÷ 2 + (bHWsinθ) ÷ 2 = (LHWsinθ) ÷ 2. When the water surface is the plane going through AMN its height above the aquarium floor is constant in the infinitesimal band shown according to the function h(x) = H − Hx ÷ (2a+b). For x < a the length of the band is w(x) = Wx ÷ a; for x > a the length is w(x) = W(L − x) ÷ b. The infinitesimal volume element whose top is the band has a volume of dVol = h(x)w(x)sinθdx. Integrate this from x = 0 to L to give a water volume of (HWLsinθ) ÷ 3 when the aquarium is tipped. This is exactly two thirds of the full volume and produces 8 gallons in one step while ignoring the 7-gallon aquarium entirely. As long as the aquarium is a cylinder of arbitrary quadrilateral cross section the aquarium's cross sectional shape doesn't matter!

6. The Persistent Snail

(a) During the nth minute the band is 100n long and the snail travels a fraction of the band equal to $1/100n$. The snail eventually arrives at the far end when $1/100 + 1/200 + 1/300 + \ldots + 1/100t$ exceeds or equals 1. That happens for $t = 1.509269 \times 10^{43}$ minutes.

(b) When the snail has traversed 99 percent of the band the 100 foot stretch causes the far end to move 1 foot farther from the snail. This 1 foot is overcome by the snail's normal progress during that minute and after this time the snail continues to get closer to the far end. This happens when $1 + 1/2 + 1/3 + \ldots + 1/u = 99$, making $u = 5.5522899 \times 10^{42}$. Note that $t \div u \approx e$, where $e = 2.7182818\ldots$

7. Shortcut Geometry

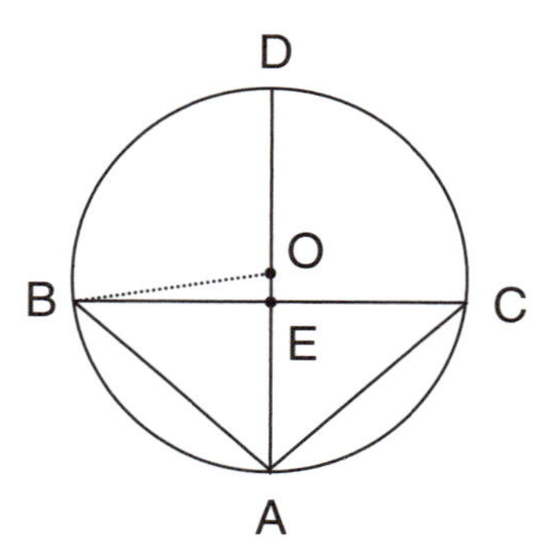

The radius of the circle isn't given in the problem so the time-pressured student solves it faster by assuming the radius doesn't matter and uses the special case shown where AD is the diameter of the circle. Then $BE^2 = CE^2 = 144 - 64 = 80$. Products of the line segments of intersecting chords are always equal, so $AE \times ED = BE \times CE = 80$ gives $ED = 10$ and $AD = 18$. Proving the result is independent of the radius can be tackled by the student at his leisure.

8. Shoelace Clocks

(a) Burn lace 2 from both ends and lace 1 from one end. When lace 2 is consumed light the other end of lace 1. The final time interval for lace 1 is 31 minutes.

(b) Burn lace 2 from both ends and lace 3 from one end, reducing lace 3 to 12 minutes when lace 2 is consumed. Burn lace 1 from one end and lace 3 from both ends, reducing lace 1 to 122 minutes when lace 3 is consumed. Burn lace 1 from both ends for a 61-minute interval.

(c) Burn lace 1 from one end and lace 2 from both ends, reducing lace 1 to $e - \sqrt{2} \div 2$ hours when lace 2 is consumed. Burn this lace at both ends for an interval of 1.0055875...hours.

(d) Light the first lace at two internal places and one end to start with 5 burning points. Maintain 5 burning points at all times to get 12 minutes when the lace is consumed. It's fair to snuff out an end and simultaneously light a middle point to accomplish this. Twelve burning points can be maintained on the second lace to produce an additional 5-minute interval when it is consumed. This method clearly generalizes.

9. Minimum Hits

One hit suffices. Each time you serve you swing and miss, each time your opponent serves it's a double fault. You carry on this way until you serve at 6–5 or 7–6 in the tiebreak at which point you serve an ace to win the set.

10. The Three Switches

(a) Turn switches off for 10 minutes. Leave S1 off. Turn on S2 for 10 minutes and then off. Turn on S3 and climb the stairs to inspect the lamp. Unlit with a cold bulb gives S1; Unlit with a warm bulb gives S2; a lit bulb gives S3.

(b) Wait 10 minutes. Leave S1 alone. Change S2 for 10 minutes; then change it back. Change S3 and go upstairs. Unlit with a cold bulb gives S1. Unlit with a warm bulb gives S2. A lit bulb gives S3.

(c) Wait 10 minutes. Leave S1 alone. Change S2 for 10 minutes; then change it back. Wait 1 minute. Change S3 and go upstairs. Unlit with a cold bulb or lit with a hot bulb and warm lamp socket gives S1. Unlit with a warm bulb or lit with a hot bulb and a lukewarm lamp socket gives S2. Lit with a warm bulb or unlit with a hot bulb gives S3. The approach for (c) uses the fact that a 100-watt bulb takes 30 to 45 seconds to heat up or cool down to a point where it can be held comfortably. The lamp socket needs 3 to 5 minutes to get fully warm. It takes about 20 seconds to get to the third floor.

11. Transit Time

The trick here is that 1 knot = 1 nautical mile per hour, so that 1 knot per hour is a constant acceleration of $a = 1$ nmi per hour per hour; $2244.5 = at^2 \div 2$ gives $t = 67$ hours.

12. Which Circle Is Larger?

The two have equal areas. On the plane the surface area is πr^2. On a sphere of radius R the area of a spherical cap of half-angle θ is $2\pi(1 - \cos\theta)R^2$ (θ is shown in the figure at right). Also from the figure, $\cos\theta = 1 - 2\sin^2(\theta \div 2) = 1 - r^2 \div 2R^2$. Putting these together gives the area of the spherical cap also as πr^2.

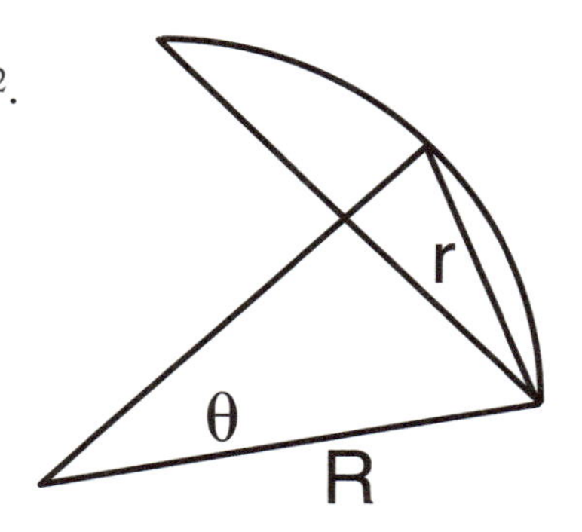

13. Modest Hexomino

The figure shows a way to cover any fraction of the hexomino arbitrarily close to 100%. Uncovered parts in the figure can be made arbitrarily small by making the neck successively thinner.

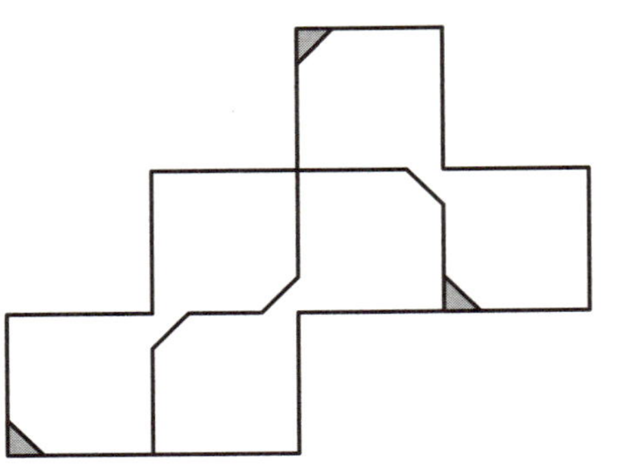

14. Logical Question

Each triangle can be broken into two identical right triangles with sides of length 3, 4, and 5. Therefore their areas are equal and the correct answer to the logical question is "No."

15. The Longest Month

October. It's one hour longer because of the end of Daylight Saving Time.

16. Outside the Box

Rotate triangle BCE 90° about B to give BAE' as shown. Angle EBE' is 90° so the area of EBE' is 2 and EE' = $2\sqrt{2}$. This makes AEE' a right triangle with area $\sqrt{2}$ so the shaded area is $2 + \sqrt{2}$.

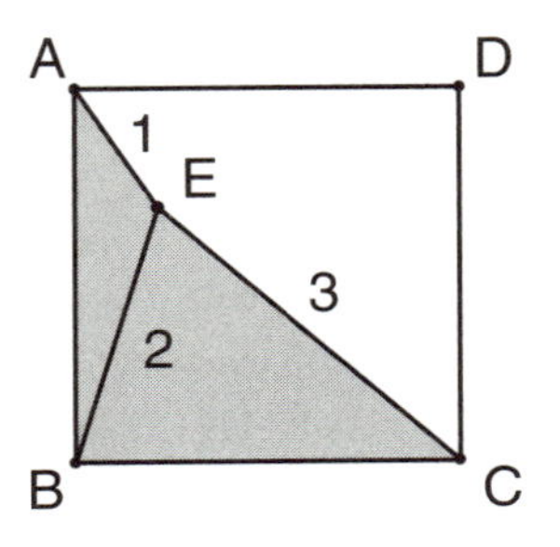

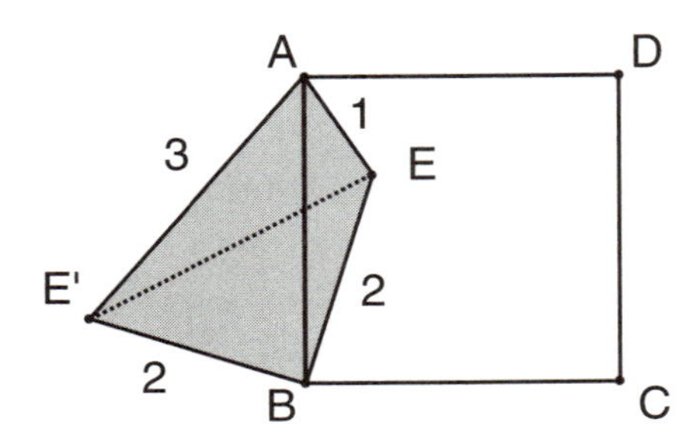

17. Happy Anniversary

The children are 5, 6, and 16 years old. To solve this, look for cases where (1) all three children are under 20 years old; (2) the younger two children have different ages; and (3) there is a product and sum which give rise to ambiguity for the ages both this year and two years ago. There is only one set of ages that accomplishes this: 5, 6, and 16. The product and sum could be achieved by 4, 8, and 15, which must have been what Smith guessed. The product and sum two years ago could have been achieved by 2, 7, and 12, which must have been what Jones guessed two years ago.

18. Magnetic Tapes

It isn't possible. Each of the initially full reels must be involved in an even number of tape transfers to end up loaded with the tape it started with. (The same applies to the initially empty reel, which must be involved in an even number of tape transfers to remain empty.) Each of the thirteen tapes must be wound or rewound an odd number of times to end up reversed. Thus the thirteen tapes have a total of an odd number of windings or rewindings to reach the goal and there is a contradiction.

19. The Upside-Down Watch

I arrived 1 hour and 40 minutes early, at 10:21. The possible times that could have been read incorrectly upside down are 10:11, 10:21, 11:01, 11:21, 12:01, and 12:11. Misreading 11:01 for 10:11 or 12:11 for 11:21 would mean that I had arrived 50 minutes early. Since we know the solution is unique I must have misread the time as 12:01 when it was actually 10:21.

20. Tom, Dick, and Harry

Ask Tom "Is Dick less likely than Harry to tell the truth?" From his answer, Tom says A is less reliable than B. Next you ask A "Between B and Tom, is Tom less likely to tell the truth?" Those indicated as less reliable in the answers to these two questions are the Knight and Knave in some order; the other is the Citizen. Now ask either of the non-Citizens, "Does two plus two equal five?"

21. Blind Logic

A doesn't see four 11's, so B doesn't see three 11's on C, D, and E, so C doesn't see two 11's on D and E, so D doesn't see any 11's—so E must have a 7.

22. Bananas and Monkeys

From the wording of the problem it is clear that from any point on a monkey's path his journey forward and backward is uniquely determined. Thus two monkeys will never traverse the same ladder segment, nor traverse the same rope in the same direction; if they did, they could be backtracked to the same starting point. Similarly no monkey can get caught in a looping path because no monkey starts in such a looping path. Therefore each monkey ends at the top of a ladder never visited by a prior monkey and gets a banana. The solution generalizes to any number of ladders.

23. Odd Logic

The number on A's hat is 15, and the number on B's hat is 17. The table at the top of the next page shows a corridor of possibilities that are successively eliminated by the statements until the cell containing the star is finally reached after the eighth set of statements. Clearly the approach generalizes.

What A sees

What B sees	1	3	5	7	9	11	13	15	17	19
1	A1	B1								
3		A2	B2							
5			A3	B3						
7				A4	B4					
9					A5	B5				
11						A6	B6			
13							A7	B7		
15								A8	★	
17										
19										

24. The Chameleons

Let the numbers of green, brown, and gray chameleons be x, y, and z. Consider the remainders left over when the quantities $x - y$, $x - z$, and $y - z$ are divided by three. By the problem conditions these remainders won't change as a result of any meetings. Thus a pair of colors can be eliminated only if the numbers of chameleons with these colors differ by a multiple of three. For the first set of numbers all chameleons will be gray after the following sequence of changes: 13, 19, 23; 15, 18, 22; 17, 17, 21; and then pairs of green and brown chameleons turn gray until we reach 0, 0, 55. For the second set of numbers no pair of numbers differ by a multiple of 3, so the chameleons can never all be the same color.

25. It's All Relative

The woman can be the man's daughter-in-law, spouse's aunt (spouse's mother's sibling's wife), or niece or nephew's wife (spouse's sister's child's wife). Or, equivalently, the man can be the woman's father-in-law, niece or nephew's husband (spouse's sister's child's husband), or aunt's husband (spouse's mother's sibling's husband).

26. Tennis Aces 1
You are ahead 2–1 in the tiebreak game. You must have been serving at 5–6, 0–40. Five aces got us to a tiebreak, I served one point and then you served the next two points in the tiebreak.

27. Tennis Aces 2
Roddick is ahead 1–0 in the fourth set but down two sets to one. At 5–4 (or 5–0) in the 3rd set tiebreak game he served two aces to win the set. Then he served the first game of the 4th set with four more aces.

28. Logical Hats 1
Cases with certain proportions of the numbers on A:B:C are successively eliminated by the logicians' statements:
A1: Eliminates 2:1:1.
B1: Eliminates 1:2:1 and 2:3:1. Since 35 is not divisible by 3, the proportions 2:1:1 and 1:2:1 eliminated by A1 and B1 give no information. Because 35 is divisible by 5 the eliminated proportion 2:3:1 does eliminate 14:21:7 and C must have seen the numbers 14 and 21 on A and B and knows he has 35 on his hat. This problem generalizes so that further statements by A, B, and C that they don't know their numbers eliminate the following proportions:
C1: 1:1:2, 1:2:3, 2:1:3, 2:3:5.
A2: 3:2:1, 4:3:1, 3:1:2, 5:2:3, 4:1:3, 8:3:5.
B2: 1:3:2, 1:4:3, 2:5:3, 2:7:5, 3:4:1, 4:5:1, 3:5:2, 5:8:3, 4:7:3, 8:13:5.
C2: 3:2:5, 4:3:7, 3:1:4, 5:2:7, 4:1:5, 8:3:11, 1:3:4, 1:4:5, 2:5:7, 2:7:9, 3:4:7, 4:5:9, 3:5:8, 5:8:13, 4:7:11, 8:13:21.
A3: 5:3:2, 7:4:3, 8:5:3, 12:7:5, 5:4:1, 6:5:1, 7:5:2, 11:8:3, 10:7:3, 18:13:5, 7:2:5, 10:3:7, 5:1:4, 9:2:7, 6:1:5, 14:3:11, 7:3:4, 9:4:5, 12:5:7, 16:7:9, 11:4:7, 14:5:9, 13:5:8, 21:8:13, 18:7:11, 34:13:21.

B3: 3:8:5, 4:11:7, 3:7:4, 5:12:7, 4:9:5, 8:19:11, 1:5:4, 1:6:5, 2:9:7, 2:11:9, 3:10:7, 4:13:9, 3:11:8, 5:18:13, 4:15:11, 8:29:21. 5:7:2, 7:10:3, 8:11:3, 12:17:5, 5:6:1, 6:7:1, 7:9:2, 11:14:3, 10:13:3, 18:23:5, 7:12:5, 10:17:7, 5:9:4, 9:16:7, 6:11:5, 14:25:11, 7:11:4, 9:14:5, 12:19:7, 16:25:9, 11:18:7, 14:23:9, 13:21:8, 21:34:13, 18:29:11, 34:55:21.

29. Logical Hats 2

Logician A's hat has 135 on it, and B's has 136. A sees 136 on B and knows whatever product B sees (1×135, 2×134, 3×133, etc.) he can't know his number. B sees 135 on A and knows he may have $3 + 45$, $5 + 27$, $9 + 15$, or $1 + 135$. The first three possibilities are eliminated since they are equal to one more than the primes 47, 31, and 23, respectively (which would make A's statement untrue, since it would be possible for B to know his number if A's unknown number were prime) so B deduces his number is 136.

30. The Sum of Squares

The numbers are 6 and 8. A sees 100 and B sees 14. The table below (showing sums of squares) enumerates all the possible pairs assuming that $x \geq y$. A1 eliminates all blank cells; B1 eliminates italic cells; A2 eliminates bold cells and B2 eliminates gray cells. Of what remains, (6,7) is eliminated since A's number is even, leaving (6,8) as the lowest possible.

	0	1	2	3	4	5	6	7	8	9	10	11	12	13	14	15	16
0						*25*					100			169		225	
1								*50*	*65*				145	170			
2										85		125			200		260
3					*25*							130			205		265
4								**65**						185			
5						**50**					125		169		221	250	
6								85	100					205			
7										130		170					
8										145		185			260	289	
9													225	250			
10											200	221					

31. Bell-Ringing Logic

A, B, and C have 2, 3, and 3 on their hats. Each round serves to eliminate one or more of the 25 initially possible triples:

R1: Eliminate 008, 017, 026, 035, and 044.
R2: Eliminate 007.
R3: 006.
R4: 016.
R5: 015 and 116.
R6: 025.
R7: 024 and 125.
R8: 034 and 115.
R9: 033, 114, and 134.
R10: 124.
R11: 123 and 224.
R12: 133.
R13: At this stage A knows he has a 2, announces it and wins the prize. B and C are each undecided between 2 and 3 on their hats.

32. Blind Bell-Ringing Logic

B has 0 on his hat. The others have 2, 2, and 2 or 2, 2, and 3 in some order. In listing all quadruples they have to be distinguished according to which digit is on the blind logician (B). For example the quadruple 0025 appears in the list as **0**025, 00**2**5, and 002**5** where the bold and underlined digit indicates it's on B. Each round serves to eliminate one or more of the 95 initially possible quadruples:

R1: Eliminate **0**008, 000**8**, **0**017, 00**1**7, 001**7**, **0**026, 00**2**6, 002**6**, **0**035, 00**3**5, 003**5**, **0**044, 00**4**4, 0**1**16, 011**6**, 0**1**25, 01**2**5, 012**5**, 0**1**34, 01**3**4, 013**4**, 0**2**24, 022**4**, 0**2**33, and 02**3**3.
R2: **0**007, 000**7**, 00**1**6, 001**6**, 00**2**5, 002**5**, 00**3**4, and 003**4**.
R3: **0**006, 000**6**, 00**1**5, 001**5**, 00**2**4, 002**4**, and 00**3**3.
R4: **0**016, 0**1**15, 011**5**, 01**2**4, 012**4**, and 01**3**3.

R5: 0015, 0114, 0114, 0123, 0123, 0116, 1115, 1115, 1124, 1124, and 1133.
R6: 0025, 0124, 0223, 0223, and 1114.
R7: 0024, 0123, 0125, 0222, 1124, 1223, and 1223.
R8: 0034, 0115, 0133, 1114, 1123 and 1123.
R9: 0033, 0114, 0134, 1113, 1113, 1122, and 1133.
R10: 0124, 1123, and 1222.
R11: 0224, 0123, 1122, 1223, and 2222.
R12: 0133, and 1222.
After the 12th round the only quadruples remaining are 0222, 0223, and 0233, from which B deduces he must wear 0. If the players wore 0233 then the logician wearing the 2 would also know his number in contradiction to the statement that B took the entire prize. Thus the others were wearing 2, 2, and 2 or 2, 2, and 3 in some order.

33. Tennis Tournament

Since two players participate in each set there are 19 sets played in all. The only way for Roddick to play just 9 sets is to play in the even numbered sets and lose each time. Thus Nadal and Federer competed in set number 13.

34. Overlapping Cookies

Let the area of a small circle equal C and let X and Y stand for the areas of X and Y. The area covered by the smaller circles is 4C − 4Y. It is also equal to the area of the large circle less 4X. Since the diameter of the large circle is twice the diameter of the small circles, the area of the large circle is 4C. Thus 4C − 4X = 4C − 4Y and X = Y.

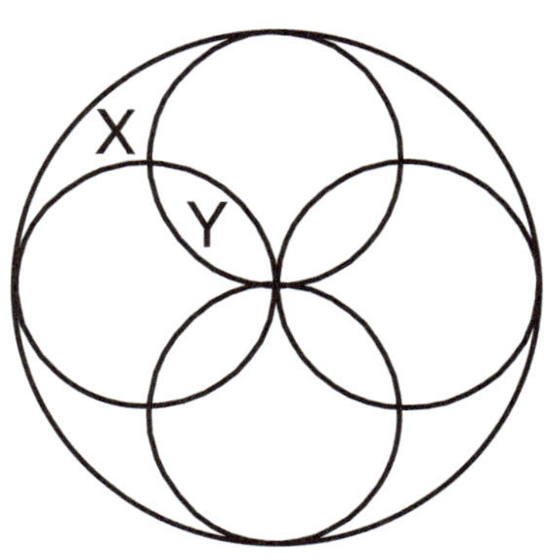

35. Measuring the North Pole

Let h = the height of the Pole, r = the earth's radius of curvature, and d = stretch in the band due to the Pole. Then (using ≈ to indicate close approximation):

(1) $d = 2r(\tan A - A) = 2r(A^3 \div 3 + 2A^5 \div 15 + \ldots) \approx 2r(A^3 \div 3)$.

(2) $h = r(\sec A - 1) \approx rA^2 \div 2 \approx (9d^2r \div 32)^{1/3} \approx 180.749192$ feet.

A more exact calculation gives $h = 180.749659$ ft.

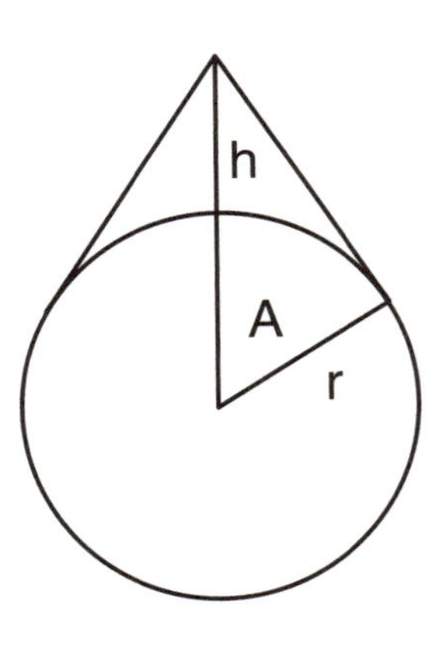

36. Tiling the Triangle

The figures show three solutions.

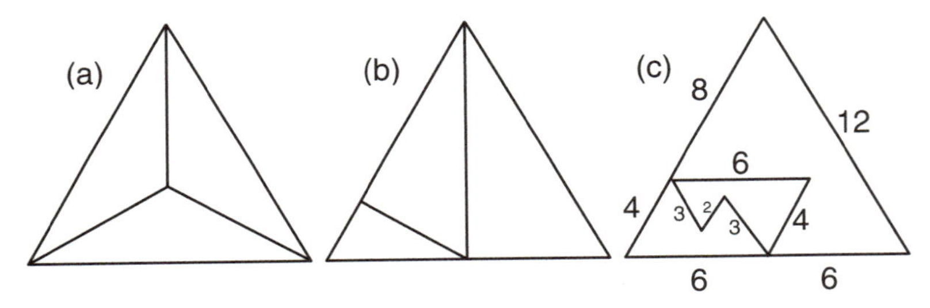

37. Reflected Pyramids?

XY needn't be perpendicular to the plane ABCD. Consider two circles of different sizes shown at right, each with a radius < 1. They define the four tangent lines drawn to produce ABCD as shown. Place X and Y out of the plane of ABCD directly above the centers of the circles so that all triangular altitudes on the two pyramids are unit. Clearly XY is not perpendicular to the plane ABCD.

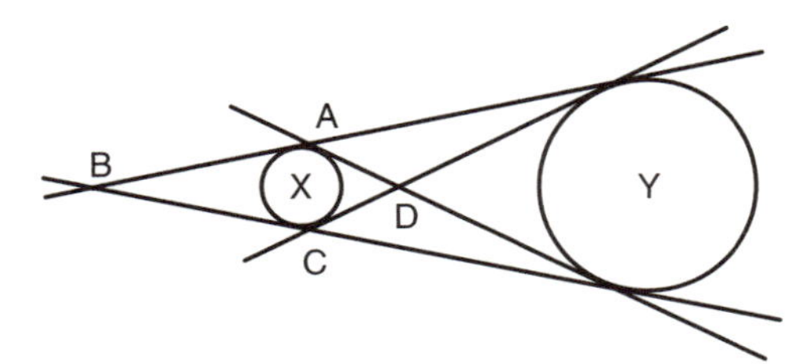

38. Divide by Three

Here are the 6 known solutions.

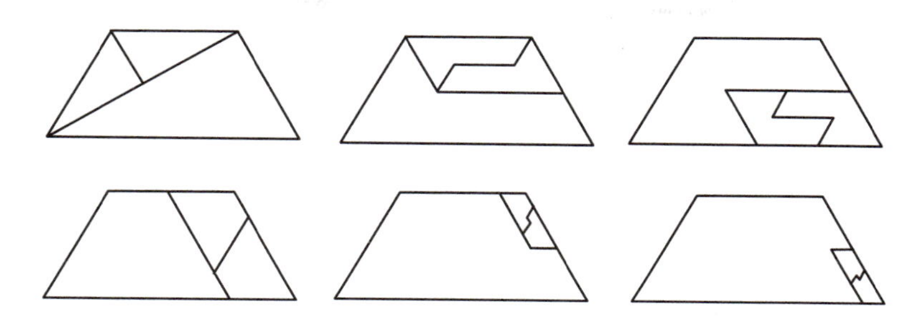

39. Divide by Four

Here are the 7 known solutions.

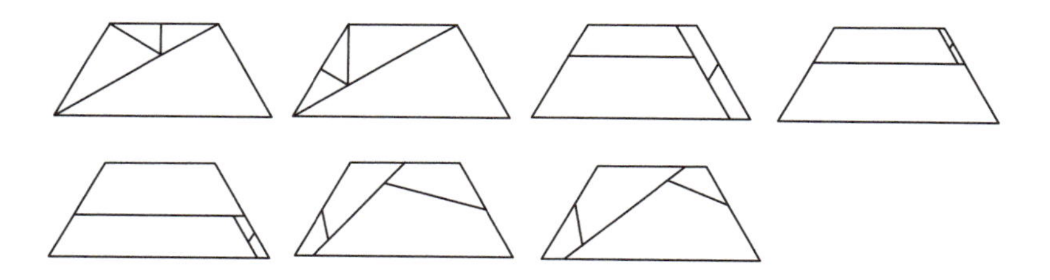

40. Squeeze In

For n = 164 there is just enough room for 329 circles if the circles are packed as shown below. There are 7 circles on each end with 105 sets of 3 circles in the middle. The smallest rectangle found so far containing 329 circles has 13 circles on each side of 101 sets of 3 and is 2 by 163.9973967....

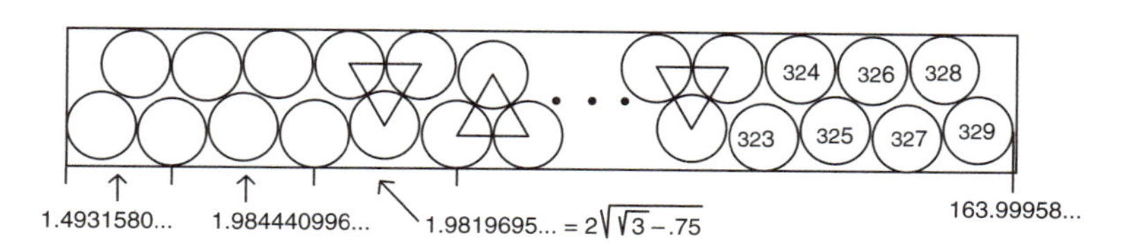

41. The Billiard Ball

Consider the ball rotating about the axis AC without slipping as shown at right. The cone BCB' rolls along the plane. The distance from B to AC is $1/\sqrt{2}$ of the distance BC. Thus B will be in contact with the horizontal plane again after $1/\sqrt{2}$ rotations about AC have occurred. Point P will return to the top again as well.

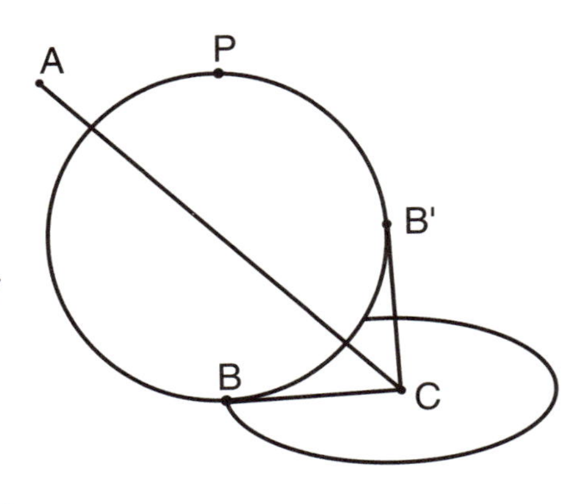

For a full rotation of the ball on the circle the point P will execute $\sqrt{2}$ rotations about the axis. Relative to the center point of the ball, point P will have coordinates $P = [-(r/\sqrt{2})\sin\theta, r(\cos\theta - 1) \div 2, r(\cos\theta + 1) \div 2]$, where $\theta = 2\pi\sqrt{2}$. For $r = 1$, $P = (-.3629497, -.9291081, .0708919)$; its initial position was $(0, 0, 1)$.

42. Similar Triangles

The figures show the 19 known solutions. Numbers in the triangles give relative areas.

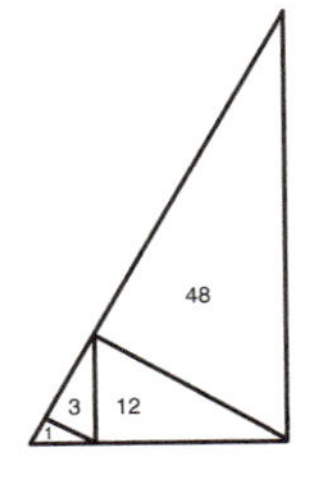

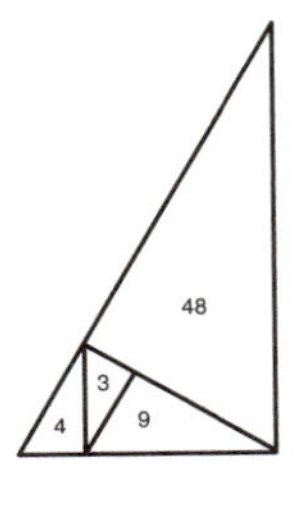

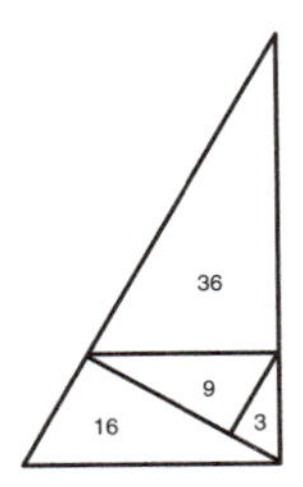

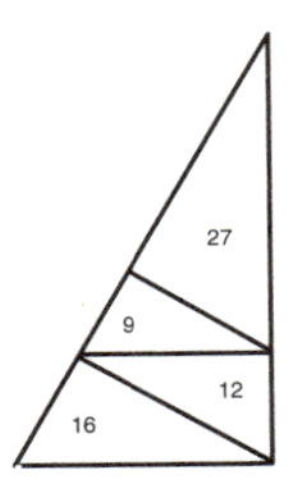

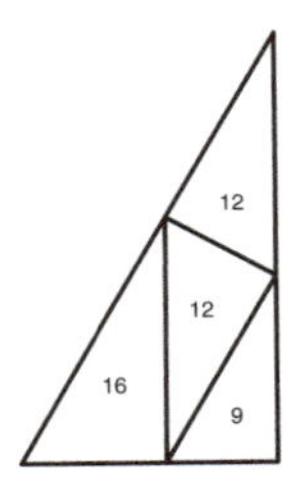

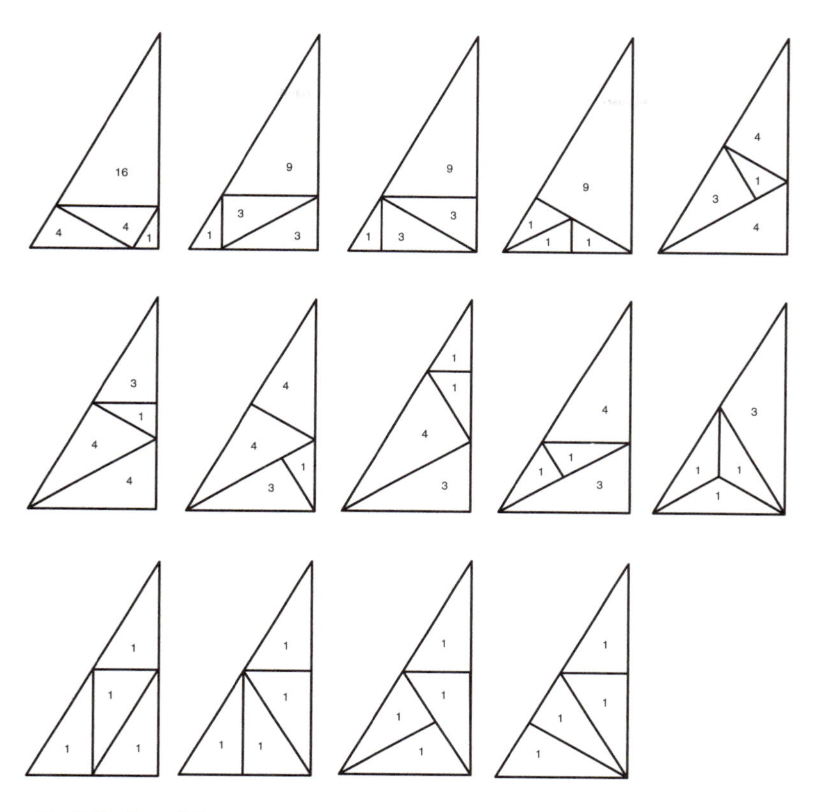

43. Modest Tromino

The figure shows a covering of 95.98402%, the best achieved so far.

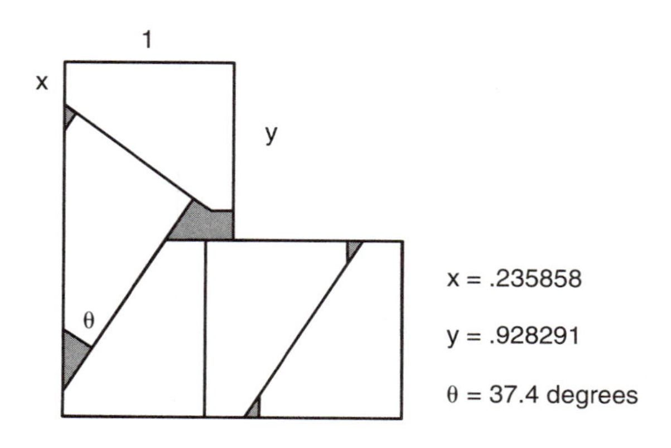

44. Modest Tetromino

The figure shows a covering of 98.1917%, the best achieved so far.

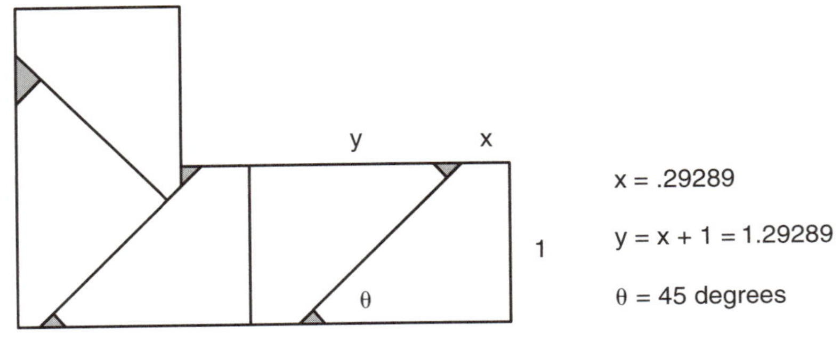

45. Seven Points

All lengths shown are 1 meter.

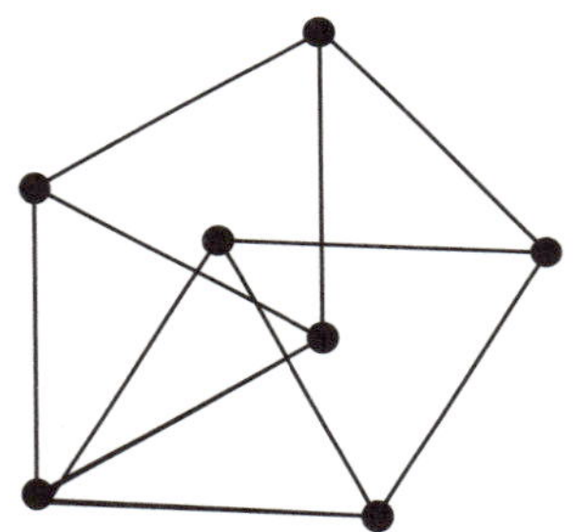

46. Where on Earth?

He can be anywhere less than 10 miles from the North Pole.

47. Linoleum Puzzle

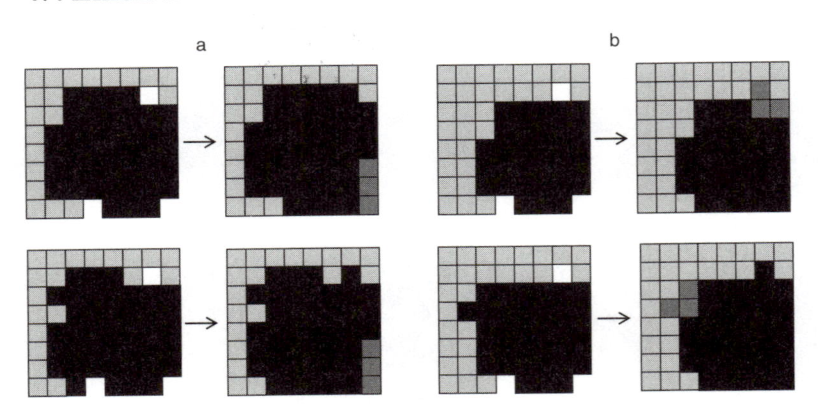

48. Counting Triangles

(a) ACK, ADJ, ADK, AEL, AGJ, AJL, CGJ, BCL, BDL, BEK, BEL, EHL.

(b) ACI, AEG, AEJ, AGI, BCH, BCI, BCJ, BFK, BHJ, CEL, CFL, CHJ, CIK, CIL, CJL, DFK, DFL, DIK, EGL.

(c) ACJ, ACM, AGI, AGJ, AHI, AHM, BDK, BDN, BEK, BGI, BHJ, BHK, BIK, BLN, CEJ, CEK, CHJ, CHL, DEN, DFK, DFL, DIM, DKL, EGM, EKL, EKM, FHN, FIN, FKL, FLN, GIM, GIN, HJN.

(d) ACJ, CAN, AGI, AGJ, AGP, AHI, AHN, AJP, ALO, BDK, BDO, BEK, BEP, BGI, BHJ, BHK, BIK, BLM, BMO, CEJ, CEK, CFL,CHJ, CHM, CIP, CJL, CNP, DEL, DEO, DFK, DFL, DFM, DIN, DKM, EGN, EKM, EKN, ELM, FHO, FIO, FKM, FLN, FLO, FMO, GIN, GIO, GJP, GLN, GNP, HIP, HJO, HJP.

49. Dissected Square 1

A solution with nine rectangles is shown.

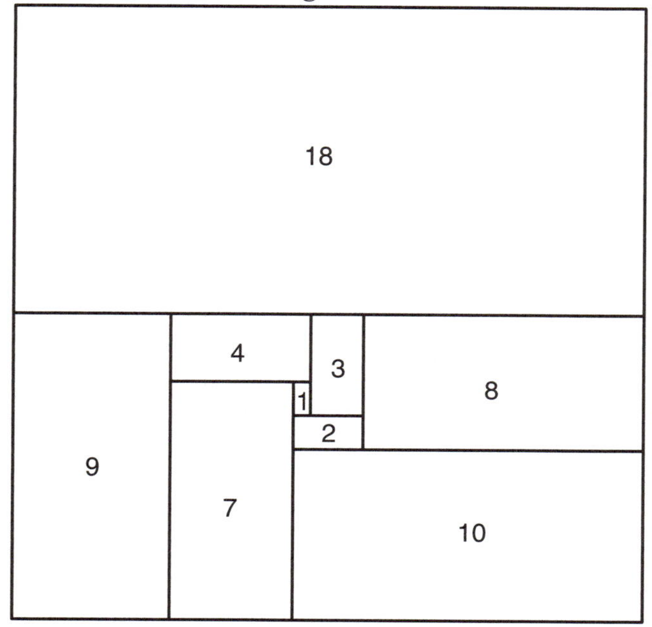

50. Dissected Square 2

The figure shows the best known solution, which uses eight triangles.

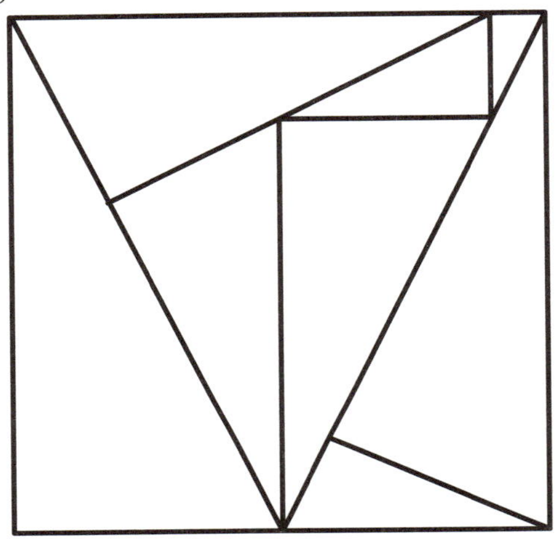

51. Irrational Pouring

Two approaches are available: (1) Repeatedly fill the e-container and pour into the π-container until it is full. Empty the π-container each time and continue. When the π-container has been emptied $b - 1$ times and the e-container has been filled a times and partially emptied into the π-container to fill it, the volume remaining in the e-container will be $ae - b\pi$ and the total number of transfers will be $n = 2(a + b - 1)$. (2) Take the same steps as in (1) but interchange the roles of the containers. In this case the volume remaining in the π-container will be $a\pi - be$ and the total number of transfers will be $n = 2(a + b - 1)$. We're looking for the smallest integer values of a and b so that $n = 2(a + b - 1)$ is minimum and $.99 < ae - b\pi < 1.01$ or $.99 < a\pi - be < 1.01$. By numerical search we find $73\pi - 84e = 1.00059$ and $57e - 49\pi = 1.004024$ are the smallest values satisfying the above conditions. Thus $a = 57$, $b = 49$ and $n = 210$ transfers is the minimum.

52. What Time Is It?

It's 8:24. If the hour hand is exactly on a second mark then the second hand will always be on the 12. For the second hand to be 18 second marks ahead of the hour hand the hour hand must be at the 42nd second mark and the time is 8:24.

53. Maximum Ratio

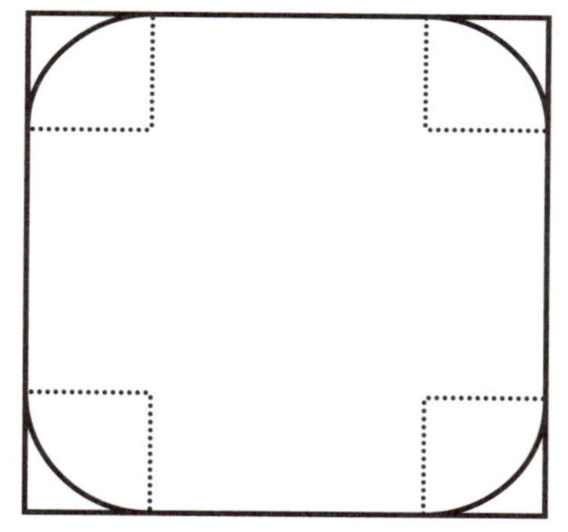

(a) Imagine an expanding soap bubble in the unit square. Its form will take on the shape of four quarter circles of radius r at the corners as shown in the figure. The area of such a shape is $A = 1 - 4r^2 + \pi r^2$. The perimeter of the shape is $P = 4 - 8r + 2\pi r$. The ratio A/P takes on a maximum for A/P = r when $r = 1 \div (2 + \sqrt{\pi}) = .265079\ldots$. Note that the ratio A/P is only .25 for the entire square or for a circle inscribed in the square.

(b) The corresponding problem of maximizing the ratio of volume to surface area within the unit cube remains unsolved. The ratio is 1/6 for the entire cube or an inscribed sphere. Expanding a soap bubble inside a cube produces nonspherical surfaces as soon as the bubble interacts with the faces of the cube. A volume with spherical octants at the 8 vertices and quarter circular cylinders on the 12 edges is not optimal but can achieve a ratio of .1852964 for a radius of .25848. Can readers make improvements?

54. Coffee Break

There are solutions for n = 1–6, 8, 9, 10, 12, 13, 15, 16, 20, 25, 50 and 100.

Notation: (A, B, fA, fB) indicates A cups and fA of a packet in the 5 cup mug and B cups and fB of a packet in the 3 cup mug.

n = 20 (2 steps): (5, 0, 0, 0), (5, 0, 1, 0).
n = 50 (3 steps): (5, 0, 0, 0), (2, 3, 0, 0), (2, 3, 1, 0).
n = 8 (4 steps): (5, 0, 0, 0), (5, 0, 1, 0), (2, 3, .4, .6), (5, 3, .4, .6).
n = 100 (5 steps): (0, 3, 0, 0), (3, 0, 0, 0), (3, 3, 0, 0), (5, 1, 0, 0), (5, 1, 0, 1).
n = 12 (6 steps): (5, 0, 0, 0), (5, 0, 1, 0), (2, 3, .4, .6), (0, 3, 0, .6), (3, 0, .6, 0), (5, 0, .6, 0).
n = 25 (7 steps): (5, 0, 0, 0), (2, 3, 0, 0), (2, 0, 0, 0), (0, 2, 0, 0), (5, 2, 0, 0), (4, 3, 0, 0), (4, 3, 1, 0).
n = 16 (7 steps): (0, 3, 0, 0), (3, 0, 0, 0), (3, 3, 0, 0), (3, 3, 0, 1), (5, 1, $.\overline{6}$, $.\overline{3}$), (3, 3, .4, .6), (5, 1, .8, .2).
n = 4 (10 steps): (0, 3, 0, 0), (3, 0, 0, 0), (3, 3, 0, 0), (3, 3, 0, 1), (5, 1, $.\overline{6}$, $.\overline{3}$), (3, 3, .4, .6), (5, 1, .8, .2), (0, 1, 0, .2), (1, 0, .2, 0), (5, 0, .2, 0).
n = 5 (10 steps): (5, 0, 0, 0), (2, 3, 0, 0), (2, 0, 0, 0), (0, 2, 0, 0), (5, 2, 0, 0), (4, 3, 0, 0), (4, 3, 1, 0), (4, 0, 1, 0), (1, 3, .25, .75), (5, 3, .25, .75).
n = 15 (12 steps): (5, 0, 0, 0), (2, 3, 0, 0), (2, 0, 0, 0), (0, 2, 0, 0), (5, 2, 0, 0), (4, 3, 0, 0), (4, 3, 1, 0), (4, 0, 1, 0), (1, 3, .25, .75), (0, 3, 0, .75), (3, 0, .75, 0), (5, 0, .75, 0).
n = 1 (13 steps): (0, 3, 0, 0), (3, 0, 0, 0), (3, 3, 0, 0), (3, 3, 0, 1), (5, 1, $.\overline{6}$, $.\overline{3}$), (3, 3, .4, .6), (5, 1, .8, .2), (0, 1, 0, .2), (1, 0, .2, 0), (1, 3, .2, 0), (4, 0, .2, 0), (1, 3, .05, .15), (5, 3, .05, .15).

n = 2 (13 steps): (5, 0, 0, 0), (5, 0, 1, 0), (2, 3, .4, .6), (0, 3, 0, .6), (3, 0, .6, 0), (3, 3, .6, 0), (5, 1, .6, 0), (3, 3, .36, .24), (5, 1, .52, .08), (0, 1, 0, .08), (1, 0, .08, 0), (1, 3, .08, 0), (4, 0, .08, 0).
n = 10 (14 steps): (5, 0, 0, 0), (2, 3, 0, 0), (2, 0, 0, 0), (0, 2, 0, 0), (5, 2, 0, 0), (4, 3, 0, 0), (4, 3, 1, 0), (4, 0, 1, 0), (1, 3, .25, .75), (0, 3, 0, .75), (3, 0, .75, 0), (3, 3, .75, 0), (5, 1, .75, 0), (3, 3, .45, .3).
n = 6 (14 steps): (5, 0, 0, 0), (2, 3, 0, 0), (2, 0, 0, 0), (0, 2, 0, 0), (5, 2, 0, 0), (4, 3, 0, 0), (4, 3, 1, 0), (4, 0, 1, 0), (1, 3, .25, .75), (0, 3, 0, .75), (3, 0, .75, 0), (5, 0, .75, 0), (2, 3, .3, .45), (5, 3, .3, .45).
n = 3 (15 steps): (5, 0, 0, 0), (2, 3, 0, 0), (2, 0, 0, 0), (0, 2, 0, 0), (5, 2, 0, 0), (4, 3, 0, 0), (4, 3, 1, 0), (4, 0, 1, 0), (1, 3, .25, .75), (1, 0, .25, 0), (5, 0, .25, 0), (2, 3, .1, .15), (0, 3, 0, .15), (3, 0, .15, 0), (5, 0, .15, 0).
n = 9 (15 steps): (5, 0, 0, 0), (2, 3, 0, 0), (2, 0, 0, 0), (0, 2, 0, 0), (5, 2, 0, 0), (4, 3, 0, 0), (4, 3, 1, 0), (4, 0, 1, 0), (1, 3, .25, .75), (0, 3, 0, .75), (3, 0, .75, 0), (3, 3, .75, 0), (5, 1, .75, 0), (3, 3, .45, .3), (5, 3, .45, .3).
n = 13 (15 steps): (5, 0, 0, 0), (2, 3, 0, 0), (2, 0, 0, 0), (0, 2, 0, 0), (5, 2, 0, 0), (4, 3, 0, 0), (4, 3, 1, 0), (4, 0, 1, 0), (1, 3, .25, .75), (0, 3, 0, .75), (3, 0, .75, 0), (3, 3, .75, 0), (5, 1, .75, 0), (3, 3, .45, .3), (5, 1, .65, .1).

The most difficult to get is 4 cups of 3% (21 steps): (5, 0, 0, 0), (2, 3, 0, 0), (2, 0, 0, 0), (0, 2, 0, 0), (5, 2, 0, 0), (4, 3, 0, 0), (4, 0, 0, 0), (4, 0, 1, 0), (1, 3, .25, .75), (0, 3, 0, .75), (3, 0, .75, 0), (3, 3, .75, 0), (5, 1, .75, 0), (3, 3, .45, .3), (5, 1, .65, .1), (3, 3, .39, .36), (5, 1, .63, .12), (0, 1, 0, .12), (1, 0, .12, 0), (1, 3, .12, 0), (4, 0, .12, 0).

Next most difficult to get is 2 cups of 13% (17 steps): From above get to (5, 1, .65, .1). Then (5, 0, .65, 0), (2, 3, .26, .39).

55. The King's Reward

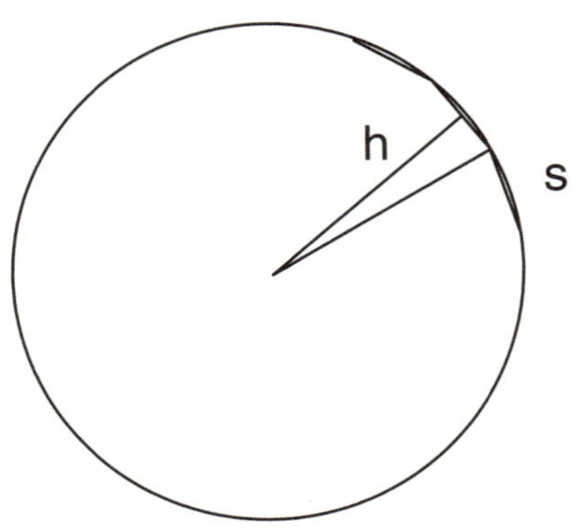

Walk a regular polygon of n sides. Its area is $A = nsh \div 2$, where s is the side length and h is the distance from the center of the polygon to the center of a side $[2h \div s = \cot(180 \div n)]$.

(a) If you travel at speed v then $s \div v = (1440 - n) \div n$ where 1440 is the number of minutes in a day. To maximize the area pick that n which maximizes $A = ns^2\cot(180 \div n) \div 4 = v^2\cot(180 \div n)(1440 - n)^2 \div (4n)$. For $n = 17$ the area is largest, with $A = 159300.1968v^2$.

(b) $s \div v = (1440 - n \div 60) \div n$.
$A = v^2\cot(180 \div n)(1440 - n\div60)^2 \div (4n)$.
For $n = 66$ the area is largest, with $A = 164635.3862v^2$. Improving the walking speed by 2% in part (a) gives $A = 165735.9247v^2$, more area than for the 1-second case.

56. The Irrational Punch

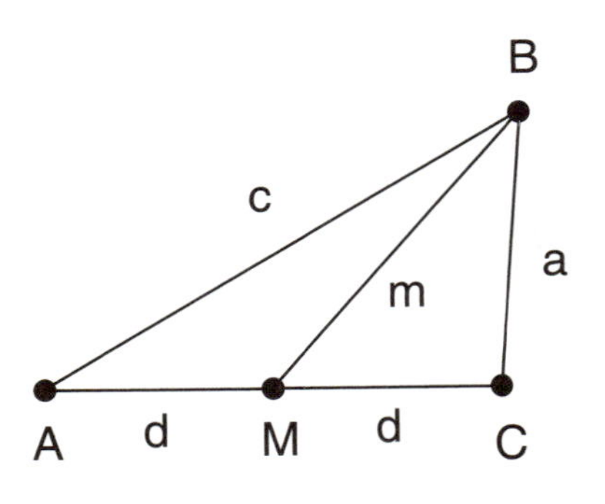

Three equally spaced points placed in a straight line do the job if the square of the spacing is irrational. In the figure, the punch has been centered on the points A, M, and C. The point B is arbitrary and m is the median of triangle ABC. It obeys the relationship $2d^2 = a^2 + c^2 - 2m^2$. Clearly, if d^2 is irrational then at least one of a, c, or m is irrational.

57. N Thirsty Customers

(a) (37, 0, 0), (31, 0, 6), (31, 6, 0), (25, 6, 6), (25, 11, 1), (36, 0, **1**), (36, 0, 0), (30, 0, 6), (30, 6, 0), (24, 6, 6), (24, 11, 1), (35, 0, **1**), (35, 0, 0), (29, 0, 6), (29, 6, 0), (23, 6, 6), (23, 11, **1**), (23, 11, 0), (23, 5, 6), (29, 5, 0), (29, 0, 5), (18, 11, 5), (18, 10, 6), (24, 10, 0), (24, 4, 6), (30, 4, 0), (30, 0, 4), (19, 11, 4), (19, 9, 6), (25, 9, 0), (25, 3, 6), (31, 3, 0) (31 steps).

(b) (37, 0, 0), (26, 11, 0), (26, **1**, 10), (26, 0, 10), (36, 0, 0), (25, 11, 0), (25, **1**, 10), (25, 0, 10), (35, 0, 0), (24, 11, 0), (24, 1, 10), (34, **1**, 0), (34, 0, 0), (23, 11, 0), (23, 1, 10), (33, **1**, 0), (33, 0, 0), (22, 11, 0), (22, **1**, 10), (22, 0, 10), (11, 11, 10), (21, 11, 0), (21, 1, 10), (31, 1, 0), (31, 0, 1), (20, 11, 1), (20, 2, 10), (30, 2, 0), (30, 0, 2), (19, 11, 2), (19, 3, 10), (29, 3, 0), (29, 0, 3), (18, 11, 3), (18, 4, 10), (28, 4, 0), (28, 0, 4), (17, 11, 4), (17, 5, 10), (27, 5, 0) (40 steps).

58. Trapping the Knight

The knight can be trapped in 15 moves as shown below.

				1				
	3							
			2		0			
		4				14		
5				15				13
		6				12		
			8		10			
	7						11	
				9				

59. Tennis Paradox

Let P(A–B) be the probability of the server winning the game when the server has A points and the receiver has B points. Let $p =$ the probability of the server winning a point and define $q = 1 - p$. Then: $P(40–40) = pP(40–30) + qP(30–40)$; $P(40–30) = p + qP(40–40)$; $P(30–40) = pP(40–40)$. This leads to $P(40–40) = p^2 \div (p^2 + q^2)$; $P(40–30) = p + p^2q \div (p^2+q^2)$. Similarly one can derive $P(40–15) = p + pq + p^2q^2 \div (p^2 + q^2)$; $P(30–15) = p^2(1 + q) + p^2q(pq + 1) \div (p^2 + q^2)$; $P(0–0) = p^4[1 + 4q + 10q^2 \div (p^2 + q^2)]$. At a score of n games to n ($n = 0$ to 5), symmetry tells us the probability of winning the set is 50% for each player. But $P(0–0) > P(30–15)$ when $8p^2 - 4p > 3$ which means the probability of winning the set from n–all, 30–15 is less than 50% when $p > .911437827$. Similarly, $P(0–0) > P(40–30)$ when $8p^3 - 4p^2 - 2p > 1$ which means the probability of winning the set from n–all, 40–30 is less than 50% when $p > .919643377607$.

60. Obtuse Triangle

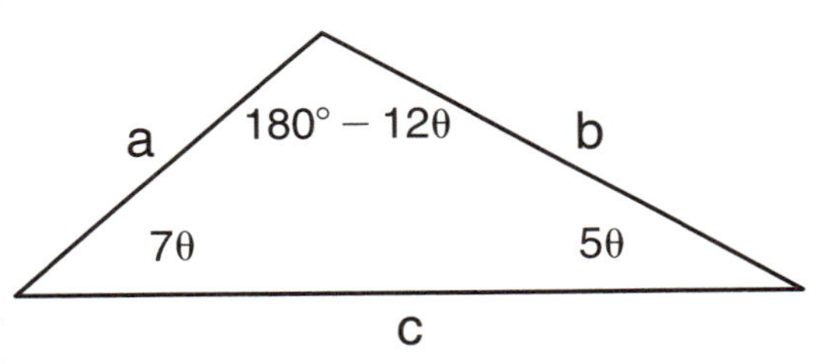

Let the acute angles be 5θ and 7θ. Since a, b, and c are integers we must have $\cos\theta = p \div q$ where p and q are integers. This follows from $\cos 5\theta$ and $\cos 7\theta$ being rational and repeated use of $\cos(A - B) = 2\cos A\cos B - \cos(A + B)$. For the remaining angle to be obtuse we need $\theta < 180° \div 24$, or 7.5°. Thus $\cos 7.5° = .99144... < p \div q < 1$. From the sine law $b \div a = \sin 7\theta \div \sin 5\theta$ and $c \div a = \sin 12\theta \div \sin 5\theta$. These can be expanded in terms of p, q, and $r = q^2 - p^2$ to give $b \div a = (7q^6 - 56rq^4 + 112r^2q^2 - 64r^3) \div (5q^6 - 20rq^4 + 16r^2q^2)$; $c \div a = p(2p^2 - q^2)(12q^8 - 256rp^2q^4 + 1024r^2p^4) \div (5q^{11} - 20rq^9 + 16r^2q^7)$. Choose $a = q^7(5q^4 - 20rq^2 + 16r^2)$

to clear all fractions. The smallest a meeting all requirements occurs for p = 116, q = 117, r = 233 giving:
a = 262315010313677624242797
b = 341879954061519855450963
c = 431074426481361267823120

61. Most Uniform Dice

The first die has 1, 2, 3, 6, 7, and 8 and the second die has 1, 1, 2, 3, 4, and 4, giving a frequency pattern of 2, 3, 4, 4, 3, 4, 3, 4, 4, 3, 2.

62. A 9-Digit Number

The number is 381654729.

63. The 1234 Quadrilateral

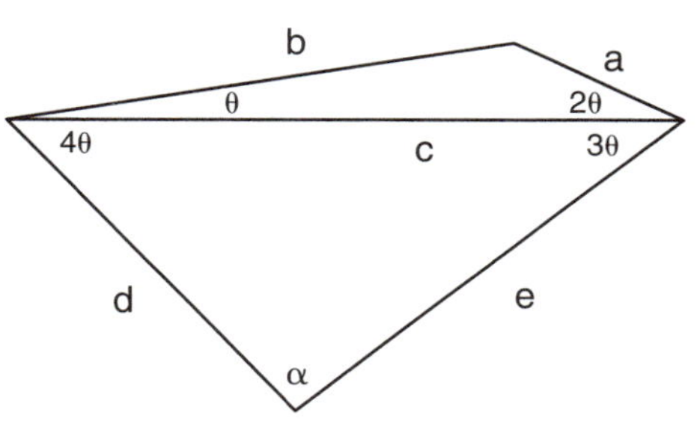

(a) a, b, and c are integers only if $\cos\theta = p \div q$, where p and q are relatively prime integers. Also $7\theta < 180°$ leads to $q > 10$. The sine law gives $b \div a = \sin 2\theta \div \sin\theta$ and $c \div a = \sin 3\theta \div \sin\theta$. These expressions can be put in terms of p and q and cleared of all fractions so that $a = kq^2$, $b = 2kpq$, and $c = k(4p^2 - q^2)$ where k is any integer constant. Similarly, $e \div d = \sin 4\theta \div \sin 3\theta$ and $c \div d = \sin 7\theta \div \sin 3\theta$. These expressions can also be put in terms of p and q by using the trigonometric expansions of multiple angles. The minimum k then can be chosen to assure all the required lengths are integer. The result is q = 12 leading to a = 1095156, b = 2007786, c = 2585785, d = 9363600, e = 9896040.
(b) When θ must be obtuse we get $q > 39$. Thus pick q = 40: a = 115208704400, b = 224656973580, c = 322872394081, d=201062560000, e=252171192000.

64. Dots on the Forehead

If you see two the same guess the opposite; otherwise, pass. This gives them a 75% chance of winning the prize.

65. Two Racers

Let the fast racer travel at speed S and the slower one at speed s. Then from the conditions of the problem $60 = (S + s) \div (S - s)$ and we get $S/s = 61/59$.

66. The 600 Sixes

Suppose there is such a number and divide it by whatever power of 100 produces an integer with units and tens digits not both zero. The possibilities are 06, 66. and 60. $100k + 6$ and $100k + 66$ both $= 2 \pmod 4$ and cannot be a square. $100k + 60 = 10 \pmod{25}$ and cannot be a square. Thus there is no such number.

67. Find All Primes

For $p = 2$ we get 8; for $p = 3$ we get the prime 17. For $p > 3$ we have $p = 6k + 1$ or $p = 6k - 1$. Then $2^p = 3n - 1$ and $p^2 = 3m + 1$, so that $2^p + p^2$ is always a multiple of 3 for $p > 3$.

68. The Wandering Knight

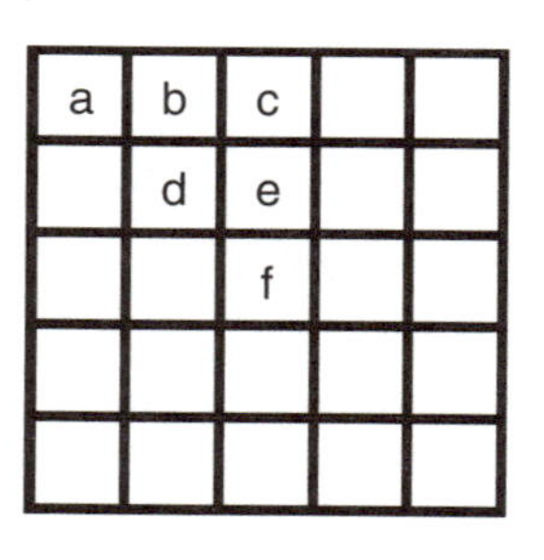

By symmetry there are six distinct occupation probabilities P_a to P_f as indicated in the figure. The probabilities for all the squares add to 1: $4P_a + 8P_b + 4P_c + 4P_d + 4P_e + P_f = 1$. The probabilities of transition from one square to another are related by $P_a = P_e \div 3$; $P_b = P_c \div 4 + P_d \div 4 + P_f \div 8$; $P_c = 2P_b \div 3 + P_e \div 3$; $P_d = 2P_b \div 3 + P_e \div 3$; $P_e = P_a + P_c \div 2 + P_d \div 2$; $P_f = 8P_b \div 3$. The solution to these equations is $P_a = 1/48$; $P_b = 1/32$; $P_c = P_d = 1/24$; $P_e = 1/16$; and $P_f = 1/12$.

69. The Pond

Let A be the surface area of the initial pond and d be its depth giving a volume of water $V = Ad$. Let the side of each cube be L. Let the water level rise amounts, x, y, and z upon the placement of the first, second, and third blocks, respectively. There are four cases possible.

(a) A very deep pond.
$V = Ad = A(d + x) - L^3 = A(d + x + y) - 2L^3 = A(d + x + y + z) - 3L^3$.
This gives $x = y = z = L^3 \div A$. This contradicts the conditions of the problem.

(b) A pond that does not completely submerge the first block but completely submerges the second block after it is placed.
$V = (A - L^2)(d + x) = A(d + x + y) - 2L^3 = A(d + x + y + z) - 3L^3$.
This gives $x = dL^2 \div (A - L^2)$, $y = 2L^3 \div A - dL^2 \div (a - L^2)$, $z = L^3 \div A$ so that $z = (x + y) \div 2$. This contradicts the conditions of the problem.

(c) A shallow pond that does not completely submerge any of the blocks.
This contradicts the conditions of the problem because the third block will cause a rise in level more than the second block.

(d) A pond that does not completely submerge the first or second blocks after they are placed but completely submerges the third block after it is placed.
$V = (A - L^2)(d + x) = (A - 2L^2)(d + x + y) = (A - 3L^2)L + A(d + x + y + z - L)$.
Equating the first and second expressions gives $Ay = L^2(d + x + 2y)$. Equating the second and third expressions gives $Az = 3L^3 - 2L^2(d + x + y)$. When $y = z$ the prior two expressions are equal so $L = d + x + 4/3y$. Then the water level above the platform is $d + x + 2y - L = 2/3y = 8$ inches.

70. Cooperative Bridge

The hand below does the job if the play goes as follows. The first five tricks are alternate spade and heart ruffs by North and West, with East underruffing North each time. North wins the sixth trick with the 10 of diamonds, East playing the 9. North next leads a heart, which West trumps with the Q. The next four tricks are won by South with the 5, 4, 3, and 2 of clubs; North and East discard all their diamonds on these tricks. The final two tricks are won by South's good diamonds.

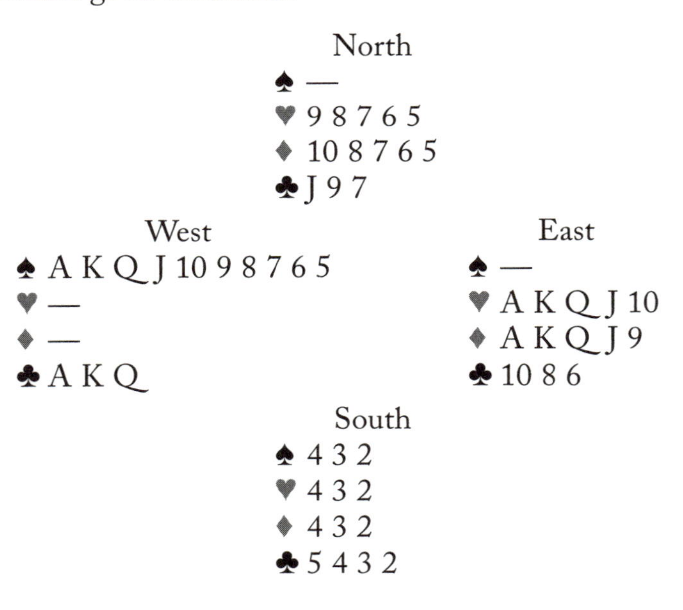

71. Cascaded Prime Triangles

(a) a = prime, $b = (a^2 - 1) \div 2$, $c = (a^2 + 1) \div 2$ = prime, $d = (c^2 - 1) \div 2$, $e = (c^2 + 1) \div 2$ = prime. Solutions occur for a = 3, 11, 19, 59, 271, 349, 521, 929, 1031, 1051, 1171, ...

(b) $a = 271$, $b = 36720$, $c = 36721$, $d = 674215920$, $e = d + 1$, $f = 227283554064939120$, $g = f + 1$. There are other solutions.

72. Eleven Coins

W1: Put 3 coins in each pan.

If the pans balance in W1 then there are true coins on the pans.

W2a: Weigh 5 of the true coins against the 5 not involved in W1 to see which is heavier.

If the pans do not balance in W1 then the excluded 5 coins are true.

W2b: Balance 3 of the true coins against the heavy group of 3 coins from W1. If they balance, the fake coin is lighter than a true coin; if not, the fake coin is heavier.

73. Four Coins and Four Weighings

W1: A + B vs. C + D
W2: A + C vs. B + D
W3: A + D vs. B + C

After the first three weighings, call the common heavy coin d and the common light coin a. Of the four coins, d is the heaviest and a is the lightest. Order the two remaining coins in a 4th weighing.

74. Eight Coins

Associate with each of the coins one of the 8 triples 000, 001, 010, 100, 110, 101, 011, and 111. The triple associated with a coin indicates the weighings that coin is involved in. Thus:

W1: D + E + F + H
W2: C + E + G + H
W3: B + F + G + H.

Calculate X = W1 ÷ 4, Y = W2 ÷ 4, Z = W3 ÷ 4. Whichever of X, Y, and Z are integers determines the false coin.

75. Stacks of Five Coins
(a) W1: $A + B + C$ vs. $3D$.

If $A + B + C = 3D$ then A, B, C, and D are true.
W2: A vs. E. $A = E$ means F is fake; $A \neq E$ means E is fake.

If $A + B + C \neq 3D$ then:
W2: $A + 3B$ vs. $2C + 2D$. If $W1 \div W2 = 1, 1/3, -1/2$, or $3/2$, then A, B, C, or D are fake.

Nine coins are involved.

(b) W1: $A + B + 3C$ vs. $D + 2E + 2F$.

If $W1 = 0$ then A, B, C, D, E, and F are true.
W2: A vs. G. $A = G$ means H is fake; $A \neq G$ means G is fake.

If W1 not equal to 0 then:
W2: $3A + B + 2C$ vs. $2D + E + 3F$. If $W1 \div W2 = 1/3, 1, 3/2, 1/2, 2$, or $2/3$ then A, B, C, D, E, or F are respectively fake.

Fourteen coins are involved.

76. Three Stacks of Six Coins
(a) W1: A vs. B.

If $A < B$ then W2: $6A + B$ vs. $6C$.
$6A + B < 6C$ means $(A, B, C) = (5, 7, 12)$
$6A + B = 6C$ means $(A, B, C) = (5, 12, 7)$
$6A + B > 6C$ means $(A, B, C) = (7, 12, 5)$

If $A > B$ then W2: $6B + A$ vs. $6C$.
$6B + A < 6C$ means $(A, B, C) = (7, 5, 12)$
$6B + A = 6C$ means $(A, B, C) = (12, 5, 7)$
$6B + A > 6C$ means $(A, B, C) = (12, 7, 5)$

(b) W1: A vs. B.

If $A < B$ then W2: $A + 6C$ vs. $4B$.
$A + 6C > 6C$ means $(A, B, C) = (6, 7, 12)$
$A + 6C = 6C$ means $(A, B, C) = (6, 12, 7)$
$A + 6C < 6C$ means $(A, B, C) = (7, 12, 6)$

If $A > B$ then W2: $6B + A$ vs. $6C$.
$B + 6C < 4A$ means $(A, B, C) = (7, 6, 12)$
$B + 6C = 4A$ means $(A, B, C) = (12, 6, 7)$
$B + 6C > 4A$ means $(A, B, C) = (12, 7, 6)$

77. Ten Stacks of Three Coins

Uniquely associate with each of the stacks one of the ten doubles 03, 30, 33, 13, 31, 12, 21, 23, 32, and 00. The double associated with a stack indicates how many coins from that stack are involved in each weighing. Then perform the following two weighings:

W1: $3B + 3C + D + 3E + F + 2G + 2H + 3I$.
W2: $3A + 3C + 3D + E + 2F + G + 3H + 2I$.

Then calculate the following:
$rA = W1 \div 18$
$rB = W2 \div 18$
$rC = (W1 - W2) \div 18$
$rD = (3W1 - W2) \div 36$
$rE = (3W2 - W1) \div 36$
$rF = (2W1 - W2) \div 18$
$rG = (2W2 - W1) \div 18$
$rH = (3W1 - 2W2) \div 18$
$rI = (3W2 - 2W1) \div 18$.

Those quantities that are integer determine the false stack.

78. N Stacks of Seven Coins

(a) $N = 512$ is the largest. Associate one of the 512 triples from 000 to 777 uniquely to each of the 512 stacks. The triple associated with a stack indicates how many coins from that stack are involved in each weighing. Each weighing involves 1792 coins. The values of $|W1 - 1792T| \div T$, $|W2 - 1792T| \div T$ and $|W3 - 1792T| \div T$ give the triple associated with the stack of fakes.

(b) $N = 416$ is the largest. There are 416 triples with entries 0–7 having unique proportions a:b:c. Uniquely associate one of these to each of the 416 stacks. The triple associated with a stack indicates how many coins from that stack are involved in each weighing. For one choice of the 416 triples 1495 coins are used. The proportions $|W1 - 1495T|:|W2 - 1495T|:|W3 - 1495T|$ give the triple associated with the stack of fakes.

79. Four Stacks of Twenty Coins

(a) W1: $A + B$ vs. $C + D$. Relabel so $A + B \leq C + D$
  If $A + B < C + D$ then W2: $3A + C$ vs. $B + 2D$
    If $3A + C > B + 2D$ then W3: $3B$ vs. D.
    If $3A + C \leq B + 2D$ then W3: $3A$ vs. C.
  If $A + B = C + D$ then W2: $B + 3D$ vs. $A + 3C$
    If $B + 3D > A + 3C$ then W3: $2A$ vs. D.
    If $B + 3D \leq A + 3C$ then W3: $2D$ vs. B.
At most 9 coins are used.

(b) W1: $A + B$ vs. $C + D$. Relabel so $A + B \leq C + D$
  If $A + B < C + D$ then W2: $20D$ vs. $17A + 20B + 9C$.
    If $20D > 17A + 20B + 9C$ then W3: $6B$ vs. C.
    If $20D \leq 17A + 20B + 9C$ then W3: $2A$ vs. C.
  If $A + B = C + D$ then W2: $4B + C$ vs. $2A + 3D$.
    If $4B + C > 2A + 3D$ then W3: C vs. $3D$.
    If $4B + C \leq 2A + 3D$ then W3: $2B$ vs. C.
At most 66 coins are used.

80. N Stacks of Twelve Coins

(a) There are 12675 triples with entries −12 to +12 having unique signed proportions a:b:c. Associate one of these to each of 12675 stacks. A triple associated with a stack indicates for each weighing how many coins and which side of the balance the coins go for that stack. W1 = W2 = W3 = 0 means the stack associated with 000 has the fakes. W1:W2:W3 = a:b:c identifies the fake stack as the one associated with the triple abc.

(b) There are 355425 quadruples with entries −12 to +12 having unique signed proportions a:b:c:d. Follow the approach as in (a) above to find the fakes among 355425 stacks.

Note that the method for computing triples and quadruples begins by defining G(k) as the number of k-tuples of positive integers 1 to 12 such that no two k-tuples are multiples of each other. G(0) = G(1) = 1. By the principle of inclusion and exclusion calculate:

$G(k) = 12^k - 6^k - 4^k - 2^k - 1^k - 1^k + 2^k + 1^k$, for $k = 2, 3$, and 4.

From all 12^k possibilities we've subtracted out those with multiples of 2, 3, 5, 7, and 11, and then added back the multiples of 6 or 10 that were subtracted out twice. This gives G(2) = 91, G(3) = 1447, G(4) = 19183. Then: triples = 8G(3) + 12G(2) + 6G(1) + G(0) = 12675, and quadruples = 16G(4) + 32G(3) + 24G(2) + 8G(1) + G(0) = 355425.

81. 2006 Coins

Divide the coins into 5 groups. A, B, C, and D have 501 coins; E has 2.

W1–2: A vs. B and C vs. D. Relabel so $A \leq B$ and $C \leq D$.

(1) If $A = B$ and $C = D$, either $H + L = 2T$ or H and L are in E.

W3: E1 vs. E2. Balance implies $H + L = 2T$. If no balance then

W4: $E1 + E2$ vs. 2 true coins from A determines how $H + L$ compares to 2T.

(2) If $A = B$ and $C < D$ then split each of A and C into two subsets. A1 and C1 have k coins; A2 and C2 have $501 - k$ coins where k can be any integer from 1 to 500.

W3: A1 vs. C1.

If $A1 = C1$ then H and L are in $C2 + D + E$.

W4: $C2 + D + E$ vs. a like number of coins from $A + B + C1$.

If $A1 > C1$ then H and L are in $C1 + D + E$.

W4: $C1 + D + E$ vs. a like number of coins from $A + B + C2$.

(3) If $A < B$, $C < D$ then:

W3: A vs. C.

If $A < C$ then L is in A and H is in D.

W4: $A + D$ vs. $B + C$.

If $A > C$ then L is in C and H is in B.

W4: $A + D$ vs. $B + C$.

This approach works for any N coins when $N > 7$ except for $N = 11$ or 15.

82. Four 4's Warmup

$1 = 4 \times 4 \div (4 \times 4)$; $2 = 4 \times 4 \div (4 + 4)$;
$3 = (4 + 4 + 4) \div 4$; $4 = 4 + 4 \times (4 - 4)$;
$5 = (4 + 4) \div (4 \times .4)$; $6 = 4 \times 4 - 4 \div .4$;
$7 = 4 + 4 - 4 \div 4$; $8 = 4 + 4 + 4 - 4$;
$9 = 4 \div .4 - 4 \div 4$; $10 = 4 \div .4 + 4 - 4$;
$11 = 4 \div .4 + 4 \div 4$; $12 = 4 \times (4 - 4 \div 4)$;
$13 = (4 - .4) \div .4 + 4$; $14 = 4 \times (4 - .4) - .4$;
$15 = 4 \times 4 - 4 \div 4$; $16 = 4 + 4 + 4 + 4$;
$17 = 4 \times 4 + 4 \div 4$; $18 = 4 \div .4 + 4 + 4$;
$19 = (4 + 4 - .4) \div .4$; $20 = 4 \div .4 + 4 \div .4$

83. Twenty MathDice Challenges

(a) $5 = 7 \times .7 + .1$
(b) $20 = 4 \div (.3 - .1)$
(c) $17 = 24 - 7 = (7 - .2) \div .4$
(d) $30 = 27 \div .9$
(e) $9 = 5 + 6 - 2 = 6(2 - .5) = (5 \times .6)^2$
(f) $6 = 9^{.5} \div .5$
(g) $4 = 2 \div (.7 - .2) = (.7 - .2)^{-2}$
(h) $20 = (5 - 1) \div .2 = .1^{-2} \div 5$
(i) $12 = 8 \times (2 - .5) = 8 + 2 \div .5 = (8 - 2) \div .5 = 8 + .5^{-2}$
(j) $16 = 8 \times (7 - 5) = .5^{-7} \div 8$
(k) $15 = 9^{.5} \div .2$
(l) $5 = (3 + 7) \div 2 = 2 \div (.7 - .3) = .2^{-(.3 + .7)}$
(m) $8 = 3 + 7 - 2 = 2 \times (7 - 3) = 3 \div .2 - 7 = (.7 - .2)^{-3}$
(n) $20 = 5 \times (1 + 3) = (5 - 3) \div .1 = (1 + 5) \div .3 = 3 \div .15$
(o) $3 = (5 + 4)^{.5} = 4 - 5 \div 5 = 4 - .5 - .5 =$
$5 - .4 \times 5 = 4 \div .5 - 5 = 5 - 4^{.5}$
(p) $4 = (3 - .2) \div .7$
(q) $23 = 9 \div .5 + 5 = .5^{-5} - 9$
(r) $16 = 3 + 5 + 8 = 8 \times (5 - 3) = 8 + .5^{-3}$
(s) $35 = 6^2 - 1 = (1 + 6) \div .2 = 21 \div .6$
(t) $7 = 9 - 4 \times .5 = 9 - 4^{.5} = .5^{-4} - 9 = 4 + 9^{.5} = 49^{.5}$

84. A Dozen Devilish MathDice Challenges

(a) $16 = (.2 + .3)^{-4}$

(b) $25 = 2 \times 9 + 7 = .2^{(7-9)} = (.9 - .7)^{-2}$

(c) $11 = .5^{-4} - 5$

(d) $25 = (8 - 3)^2 = (8 - 3) \div .2 = 28 - 3 = 8 \div .32$

(e) $5 = 4 + 3 - 2 = 2 \times 4 - 3 = 4 \div 2 + 3 =$
$(4 - 3) \div .2 = 3^2 - 4 = 3 \div (.2 + .4) = .2^{(3-4)}$

(f) $25 = .2^{(1-3)} = (.3 - .1)^{-2} = 3 \div .12$

(g) $16 = 8 + 2^3 = 8 \div (.2 + .3) = 32^{.8}$

(h) $3 = 9^{4^{-.5}}$

(i) $16 = 8 \times 8 \div 4 = (4 \times 8)^{.8}$

(j) $2 = 2 \times (3 - 2) = 2 \div (3 - 2) = 2^{(3-2)} =$
$3 - 2 \div 2 = 32^{.2}$

(k) $16 = .5 \times 2^5 = .5^{(-2 \div .5)} = (.5 \times .5)^{-2} = .5^{-5} \div 2 =$
$.5^{-(.5^{-2})}$

(l) $8 = 9 - 5 \div 5 = (9 - 5) \div .5 = 9 - .5 - .5 =$
$5 + 9^{.5} = .5^{-(9^{.5})}$

85. Expert MathDice Challenges

(a) $23 = (6 - 2)! - 1$

(b) $21 = \sqrt{\sqrt{\sqrt{21^8}}}$

(c) $48 = 2(3! - 2)!$

(d) $34 = (3!)(3!) - 2$

(e) $31 = 7 + (3! - 2)!$

(f) $49 = \sqrt{\sqrt{(2 + 5)^8}}$

(g) $44 = 2 \div .1 + 4!$

(h) $44 = 88 \div \sqrt{4}$

(i) $42 = 48 - (\sqrt{9})!$

(j) $35 = 4 \times 9 - .\bar{9} = \sqrt{((\sqrt{9})!)^4} - .\bar{9} = (\sqrt{9})!^{\sqrt{4}} - .\bar{9}$

(k) $57 = 9 \times 9 - 4!$

(l) $23 = 2 + 7 \div \sqrt{.\bar{1}}$

(m) $34 = 7 + 3! \div .\bar{2}$

(n) $45 = (3 + 7) \div .\bar{2}$

(o) $22 = (8 \times .5)! - 2 = (\sqrt{\sqrt{.5^{-8}}})! - 2$

(p) $26 = 5! \div 5 + \sqrt{4} = (\sqrt[-.5]{.5})! + \sqrt{4} = \sqrt{\sqrt[-.5]{.5}} + 4!$

(q) $38 = (4! - 5) \div .5$
(r) $21 = 4! - \sqrt{8 \div .\bar{8}}$
(s) $38 = 8 + 4! \div .8$
(t) $48 = 8 \times (8 - \sqrt{4}) = \sqrt{\sqrt{\sqrt{48^8}}} = (4! \div 8)! \times 8 =$

$\sqrt{(8+8)}\,! \times \sqrt{4} = \sqrt{(8+8)}\,! + 4! = \sqrt{\sqrt{8+8}} \times 4!$

(u) $55 = \sqrt{8^4} - 9 = 8^{\sqrt{4}} - 9$
(v) $44 = 99 \times .\bar{4}$

86. More Expert MathDice Challenges

(a) $55 = \sqrt{5! + 1} \div .2$
(b) $42 = 7(\sqrt{2 \div .\bar{2}})! = 7! \div (\sqrt{.2^{-2}})!$
(c) $23 = \sqrt{\sqrt{.2^{-8}}} - 2$
(d) $33 = 8 \div .\bar{2} - 3$
(e) $35 = 3! \div .\bar{2} + 8$
(f) $66 = \sqrt{.\bar{4}} \div .\overline{01}$
(g) $31 = ((\sqrt{9})! - 2)! + 7$
(h) $33 = 5 \div .\bar{5} + 4!$
(i) $43 = 4! \div .5 - 5 = .4 \times 5! - 5$
(j) $46 = .4(5! - 5)$
(k) $39 = 48 - 9 = 4! \div .8 + 9$
(l) $46 = 8 \times (\sqrt{9})! - \sqrt{4}$
(m) $61 = \sqrt{4^{(\sqrt{9})!}} - \sqrt{9} = 4^{\sqrt{9}} - \sqrt{9}$
(n) $44 = \sqrt{\sqrt{7^8}} - 5$
(o) $19 = \sqrt{3!! \div 2 + 1}$
(p) $50 = (3 + 2) \div .1 = 3! \div .12$
(q) $38 = 5! \times \sqrt{.\bar{1}} - 2$
(r) $39 = 5! - (.\bar{1})^{-2}$
(s) $31 = (7 - .\bar{1}) \div .\bar{2}$
(t) $34 = 2(4! - 7) = (4! - .2) \div .7$
(u) $34 = \sqrt[.5]{6} - 2 = \sqrt{6! \div .\bar{5}} - 2$
(v) $26 = 5(3! - .8)$
(w) $35 = 5 + \sqrt{3!! \div .8} = \sqrt{\sqrt{\sqrt{35^8}}}$
(x) $41 = 5 + \sqrt{\sqrt{(3!)^8}}$
(y) $42 = (5 + \sqrt{4})! \div 5!$
(z) $29 = (4! - .8) \div .8$

(aa) $23 = 7 + \sqrt{.5^{-8}} = 8 \div .5 + 7$
(bb) $36 = 8 \div (.\bar{7} - .\bar{5})$

87. Make 28
(a) $28 = 1 + 6 \div .\bar{2}$
(b) $28 = 5!(.\bar{3} - .1) = 3!(5 - \sqrt{.\bar{1}})$
(c) $28 = (5 - \sqrt{.\bar{1}})(\sqrt{9})!$
(d) $28 = 7 \times \sqrt{16}$
(e) $28 = 7 + 7 \div \sqrt{.\bar{1}}$
(f) $28 = \sqrt{28^2}$
(g) $28 = \sqrt{2^{3!} + 3!!}$
(h) $28 = \sqrt{\sqrt[.5]{28}}$
(i) $28 = \sqrt{6! + .5^{-6}}$
(j) $28 = .\bar{7} \times \sqrt[.5]{6} = .\bar{6} \times 7! \div 5! = .\bar{7} \times \sqrt{6! \div .\bar{5}}$
(k) $28 = 5!(.9 - .\bar{6}) = \sqrt{(\sqrt{9})!! + .5^{-6}} = \sqrt{6! + .5^{-\sqrt{9}!}}$
(l) $28 = 6(\sqrt{9})! - 8 = 8! \div (6! + (\sqrt{9})!!)$

88. The Three 4's Problem
(a) $19 = 4! - \sqrt{4} \div .4$
(b) $34 = 4! \div \sqrt{.\bar{4}} - \sqrt{4} = 4! + 4 \div .4 = \sqrt[.4]{4} + \sqrt{4}$
(c) $35 = (4! - \sqrt{.\bar{4}}) \div \sqrt{.\bar{4}}$
(d) $45 = (4! - 4) \div .\bar{4} = \sqrt{4} \div (.\bar{4} - .4)$
(e) $53 = (4! - .\bar{4}) \div .\bar{4}$
(f) $55 = (4! - \sqrt{4}) \div .4 = (4! + .\bar{4}) \div .\bar{4}$
(g) $59 = (4! - .4) \div .4$
(h) $63 = (4! + 4) \div .\bar{4}$
(i) $68 = 44 + 4! = 4 + \sqrt{\sqrt{\sqrt{4^{4!}}}}$
(j) $78 = 4! + 4! \div .\bar{4}$

89. Killer MathDice Challenges
(a) $88 = (.\bar{2} - .1) \times 3!! = .1\bar{2} \times 3!!$
(b) $75 = \sqrt{.2^{-4} \div .\bar{1}} = .2^{-\sqrt{4}} \div \sqrt{.\bar{1}}$
(c) $33 = (7 + \sqrt{.\bar{1}}) \div .\bar{2} = 7 \div .\overline{21}$
(d) $27 = 22 + 5 = (5 - 2)! \div .\bar{2} = 5^2 + 2 =$
$5 \div .2 + 2 = \sqrt[.2]{2} - 5 = \sqrt[-.5]{.2} + 2$

(e) $23 = 22 + .\bar{9} = (2 + 2)! - .\bar{9} = (2 \times 2)! - .\bar{9} =$
$(2^2)! - .\bar{9} = \sqrt[.2]{2} - 9$

(f) $40 = 4!\sqrt{2 + .\bar{7}}$

(g) $52 = .\bar{4}(5! - \sqrt{9})$

(h) $19 = 4! \div .\bar{8} - 8$

(i) $26 = 8 \div .\bar{4} + 8 = (4! - .\bar{8}) \div .\bar{8} =$
$\sqrt{(8 + 8)\,!} + \sqrt{4} = \sqrt{\sqrt{8 + 8}} + 4!$

(j) $45 = 4 \div (.\bar{8} - .8)$

(k) $66 = (9 + \sqrt{4})(\sqrt{9})! = 4!\sqrt{9} - (\sqrt{9})! = 99\sqrt{.\bar{4}}$

(l) $33 = 1 + \sqrt[.2]{2}$

(m) $29 = 4! + .2^{-1} = 4! + 1 \div .2 = \sqrt[-\sqrt{.\bar{4}}]{.\bar{1}} + 2 =$
$\sqrt{\sqrt{\sqrt{\sqrt{.\bar{1}}}}}^{\,-4!} + 2$

(n) $33 = 4 \div .\overline{12} = 4! + (\sqrt{.\bar{1}})^{-2} = 1 + \sqrt{\sqrt[.2]{4}}$

(o) $45 = 6 \div (\sqrt{.\bar{1}} - .2)$

(p) $39 = \sqrt{\sqrt{.\bar{1}^{-8}} - 7!}$

(q) $29 = 3! \div .\bar{2} + 2 = (3! - .2) \div .2 = \sqrt[.2]{2} - 3$

(r) $57 = \sqrt{(3!! + \sqrt{4}) \div .\bar{2}}$

(s) $38 = (\sqrt{9})!(\sqrt{9})! + \sqrt{4} = \sqrt{(\sqrt{9})!! + (\sqrt{9})!! + 4}$

(t) $56 = (\sqrt{9})!! \div 9 - 4! = 9(\sqrt{9})! + \sqrt{4} = (\sqrt{4^{\sqrt{9}}})! \div (\sqrt{9})!!$
$= ((\sqrt{9})! + \sqrt{4})! \div (\sqrt{9})!! = (4! \div \sqrt{9})! \div (\sqrt{9})!! =$
$(\sqrt{\sqrt{4}^{(\sqrt{9})!}})! \div (\sqrt{9})!!$

(u) $21 = (\sqrt[-.5]{.5})! - \sqrt{9} = 5! \div 5 - \sqrt{9}$

(v) $25 = .\bar{5} \div (.8 - .\bar{7}) = \sqrt{7! \div 8 - 5} = \sqrt{\sqrt{(.7 - .\bar{5})^{-8}}} =$
$\sqrt{\sqrt{5^{8!/7!}}}$

90. Mining Decision

The venture should be undertaken, since the volume in cubic spandrals can be determined using calculus as $V = (4\pi \div 3) \times 60^3 \times 1.331^{2/3} = 1094782.208$ cubic spandrals. It's noteworthy that the volume doesn't depend on the polar or equilateral radii of Alpha Lyra IV. This problem is a generalization of the classic apple core problem, which uses a sphere rather than an oblate sphere. An alternative way to solve it is to note that if it has a solution then the answer can't depend upon the polar radius of the planet (since this number was not supplied). Assuming the planet is entirely made of krypton leads to a polar radius of 60 spandrals and an equatorial radius of 66 spandrals. Then the volume of the planet follows immediately as $V = (4\pi \div 3) \times 60 \times 66^2 = 1094782.208$ cubic spandrals.

91. Excavating on Taurus

The figure shows a cross section of the torus. Let r = AB and R = CD. There are three classes of channels that can be dug on its surface:

(a) An infinity of channels with radius r are possible.

(b) Channels with radii between $R - 2r$ and R are possible.

(c) Circular channels with radius $R - r$ are possible, which are produced by the intersection of the torus and a plane shown by the slanting line.

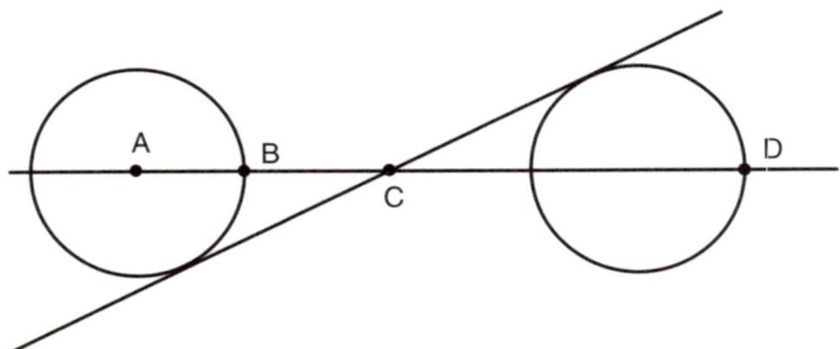

This intersection produces two circles called the Villarceau circles (shown above right). From the descriptions of the first two students they must have dug channels of type (b). The third and fourth students must have explored channels of type (a) and (c) in some order. One case gives $R - r = 25$ and $r = 30$; the other case gives $R - r = 30$

and $r = 25$. The first case is not possible because $R - 2r < 0$. Thus the second case applies and $2\pi R = 110\pi$ spandrals.

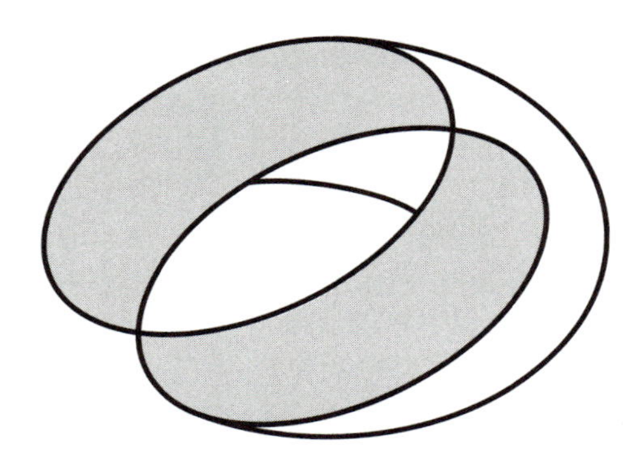

92. Life and Death on Alpha-Lyra III

(a) 45 years.

(b) 605 years.

(c) 17042641444 years.

(d) Yes; $a = 10093613546512321$ is the first of 5 in a row. $a = 49p$, $a + 1 = 2q^2$, $a + 2 = 9r$, $a + 3 = 4s$, $a + 4 = 25t$, where p through t are all primes.

(e) No. Three of the years would be even and cannot all be of the special form.

93. Humpty Dumpty

From the wording of the problem we have:

Maximum score per subject = (Qualifying score) ÷ 4

Number of subjects = (Maximum score per subject) ÷ 3

(Qualifying score) ÷ 2 = (Number of subjects) × (Number of Good Eggs − 1)

From these relations it immediately follows that there are seven Good Eggs. Trial and error then leads one to conclude there are 5 subjects with a possible score of 15 in each. The scores for the Good Eggs are (15, 14, 13, 12, 6), (15, 14, 13, 11, 7), (15, 14, 13, 10, 8), (15, 14, 12, 11, 8), (15, 14, 12, 10, 9), (15, 13, 12, 11, 9) and (14, 13, 12, 11, 10), with the last score in each set showing that achieved in arithmetic. Humpty Dumpty got 10 in arithmetic.

94. The Knights of Righteousness

Amazingly, their survival probability can be increased to 82.36195%. To do it, the knights form an assignment list in which each has a different number from 1 to 117 assigned to him. Each knight copies this list and takes it into the Room of Judgment. Upon entering the room a knight first inspects the box corresponding to his number. If his name is not in the box he next goes to the box number assigned to the name in the box and so on until he either finds his name or inspects his allowed 98 boxes. The assignment list is some permutation of the names in the boxes. This permutation, written in cyclic notation, has a longest cycle. If the longest cycle is 98 or less, then all prisoners survive. This is because each knight jumps into the cycle containing his name and will find it as the last name in that cycle. For a random permutation of n elements the probability that the longest cycle is of length k is $1/k$ if $k > n/2$. Thus the probability of survival is $1 - (1/117 - 1/116 \ldots - 1/99) = .823619\ldots$. Interestingly enough, the survival probability climbs quickly if the first few knights are known to be successful in finding their names. (98.33008% , 99.83177%, and 99.98225% after the first three knights are successful).

ACKNOWLEDGMENTS

I have attempted to identify the source of each puzzle as best I know and would be happy to hear more on the history of any of the puzzles. So many are passed along by word of mouth and adapted along the way that it's often impossible to determine the true source. Even those published in journals with problem columns often have a history predating publication. Puzzles not listed below are original by the author.

The Watermelons: I.F. Sharygin in *Quantum*, June/August 1998.

Coins in the Dark: Steven Kahan.

The Punctured Sequence, Where on Earth?: David Singmaster.

The Monster Tire: Scot Morris.

The Persistent Snail, Which Circle Is Larger?, Seven Points: From (or based on problems from) the *Pi Mu Epsilon Journal.*

Shoelace Clocks: Based on concepts introduced to me by Carl Morris and Andy Liu.

The Three Switches: Adaptation of a problem told to me by Wil Strijbos.

Outside the Box: Wei-Hwa Huang.

Magnetic Tapes: St. Petersburg Contest problem (1986).

The Upside-Down Watch, It's All Relative, The Sum of Squares, Bell-Ringing Logic, Blind Bell-Ringing Logic, What Time Is It?, Dots on the Forehead, Two Racers, Cooperative Bridge: Unknown.

Tom, Dick, and Harry; Bananas and Monkeys; The Chameleons; Reflected Pyramids?; Maximum Ratio; The King's Reward; The 600 Sixes; Find All Primes; The Wandering Knight; Four Coins and Four Weighings; The 2006 Coins: From (or based on problems from) *Crux Mathematicorum*.

Odd Logic: Based on a problem from David Gale.

Logical Hats 1: Based on a problem from Jonathan Welton.

Overlapping Cookies, The Billiard Ball, Tennis Paradox: From (or based on problems from) Puzzle Corner in *Technology Review*.

Tiling the Triangle, Divide by Three, Divide by Four, Dissected Square 1, Four 4's Warmup, Humpty Dumpty: Bob Wainwright.

Squeeze In, Trapping the Knight: Nob Yoshigahara.

Similar Triangles, Dissected Square 2: Karl Scherer.

Modest Tromino, Modest Tetromino: Collaborations with Bob Wainwright.

Linoleum Puzzle: Scott Kim.

The Irrational Punch: Ed Hess.

Eleven Coins: Andy Liu.

Four Stacks of Twenty Coins: Markus Gotz.

Mining Decision, Excavating on Taurus: Jack Porter and Joe Rathfon.

The Knights of Righteousness: Adapted from the 100 Prisoners Problem from Peter Winkler.